# Synthesis Lectures on Engineering, Science, and Technology

The focus of this series is general topics, and applications about, and for, engineers and scientists on a wide array of applications, methods and advances. Most titles cover subjects such as professional development, education, and study skills, as well as basic introductory undergraduate material and other topics appropriate for a broader and less technical audience.

Jordan Wain

# Ionising Radiation Protection

## A Primer

 Springer

Jordan Wain
Hampshire, UK

ISSN 2690-0300                ISSN 2690-0327   (electronic)
Synthesis Lectures on Engineering, Science, and Technology
ISBN 978-3-031-65524-1        ISBN 978-3-031-65525-8   (eBook)
https://doi.org/10.1007/978-3-031-65525-8

This Springer imprint is published by the registered company Springer Nature Switzerland AG
The registered company address is: Gewerbestrasse 11, 6330 Cham, Switzerland

If disposing of this product, please recycle the paper.

# Preface

Any energy production technology must be judged against the metrics of economics, security of supply, and environmental impact. Fossil fuels have many disadvantages associated with their environmental impact, and renewable sources suffer with inconsistent production; nuclear energy is the only industrially proven method to consistently supply energy without the release of large volumes of greenhouse gases. This is one of the reasons nuclear technologies have played a significant role in the energy production of many nations. Additionally, the militaries of many nations are turning to nuclear technology to power submarines. Yet nuclear energy is not perfect. Like every industry it carries risks that must be understood to be managed. With the ongoing use of nuclear energy it is essential that those involved are able to put those risks in perspective.

This book was initially written to support the Radiation Protection module of the Nuclear Advanced Course, a master's degree. The book is aimed to introduce ionising radiation protection to an undergraduate audience; however, is written assuming the reader has no specialist knowledge of the field. At the end of each chapter there are exercises to consolidate the information learnt and emphasise how it can be applied.

I have greatly benefitted from the expertise of my colleagues at the Nuclear Department, and I am honoured to express my gratitude to them.

Hampshire, UK
Jordan Wain

# Contents

1   **Introductory Nuclear Physics** ........................................... 1
    1.1   Types of Ionising Radiation ................................... 1
    1.2   Radioactive Decay ............................................ 3
    1.3   Photon Interactions .......................................... 4
    1.4   Summary ...................................................... 4
    1.5   Exercises .................................................... 5

2   **Key Units** .............................................................. 7
    2.1   Activity, Bq ................................................. 7
    2.2   Source Strength, Q ........................................... 8
    2.3   Linear Energy Transfer (LET) ................................. 8
    2.4   Absorbed Dose, $D_{abs}$ ..................................... 8
    2.5   Equivalent Dose, $H_T$ ....................................... 8
    2.6   Effective Dose, E ............................................ 9
    2.7   Chapter Summary .............................................. 10
    2.8   End of Chapter Exercises ..................................... 10

3   **Cellular and Whole-Body Responses to Radiation** ....................... 11
    3.1   Cell Recovery ................................................ 13
    3.2   Cell Death ................................................... 13
    3.3   Loss of the Ability to Replicate ............................. 15
    3.4   Cancer Formation ............................................. 15
    3.5   Hereditary Effects ........................................... 16
    3.6   Summary ...................................................... 16
    3.7   Exercises .................................................... 17

**4    Epidemiological Studies and Radiation Risk** ......................... 19
   4.1    Summary of Critical Epidemiological Studies ..................... 19
   4.2    Difficulties with Epidemiological Studies ........................ 21
   4.3    Validity of Studies ............................................ 22
   4.4    Healthy Worker Effect ........................................ 22
   4.5    Dose Limits .................................................. 23
   4.6    Summary ..................................................... 23
   4.7    Exercises ..................................................... 24

**5    Internal Radiological Exposure** ..................................... 25
   5.1    Routes of Entry ............................................... 25
   5.2    Committed Effective Dose (CED) .............................. 28
      5.2.1    The Importance of Particle Diameter for Inhalation CEDs ... 30
   5.3    Mitigating the Internal Radiation Hazard ........................ 32
   5.4    Summary ..................................................... 33
   5.5    Exercises ..................................................... 33

**6    External Radiological Exposure** ..................................... 35
   6.1    Time ......................................................... 35
   6.2    Distance ...................................................... 35
   6.3    Summary ..................................................... 37
   6.4    Exercises ..................................................... 37

**7    Background Radiation** .............................................. 39
   7.1    Primordial Background Radiation ............................... 39
      7.1.1    Oklo Reactor ......................................... 40
      7.1.2    Naturally Occurring Radioactive Material (NORM) ........ 41
   7.2    Cosmic Background Radiation .................................. 42
      7.2.1    Carbon Dating ........................................ 42
   7.3    Anthropogenic Background Radiation ........................... 43
      7.3.1    Variability of Background Radiation ..................... 43
   7.4    Summary ..................................................... 44
   7.5    Exercises ..................................................... 45

**8    Radioactive Waste Management** ..................................... 47
   8.1    Categories of Radioactive Waste ............................... 47
   8.2    Disposal Principles ............................................ 48
   8.3    Amount of Radioactive Waste .................................. 48
   8.4    Waste Activity Over Time ...................................... 48
   8.5    Disposal Options for Radioactive Waste ......................... 49
   8.6    Summary ..................................................... 49
   8.7    Exercises ..................................................... 49

**9    Radiation Detection and Instrumentation** ........................... 51
   9.1    Ion Chambers ................................................. 51
   9.2    Neutron Detectors ............................................ 52
          9.2.1   Geiger-Muller Detectors ............................... 52
   9.3    Scintillators ................................................. 53
   9.4    Semiconductors .............................................. 54
          9.4.1   Semiconductor Theory ................................ 54
   9.5    Personal Dosimeters .......................................... 55
          9.5.1   Thermoluminescent Dosimeters (TLDs) ................... 55
   9.6    Cherenkov Radiation ......................................... 56
   9.7    Calibration ................................................. 58
   9.8    Minimum Detectable Amount (MDA) ......................... 58
          9.8.1   Solid Angle, $\Omega$ ......................................... 61
   9.9    Summary ................................................... 61
   9.10   Exercises .................................................. 62

**10    Shielding** ...................................................... 63
   10.1   Alpha ..................................................... 63
   10.2   Beta ...................................................... 64
          10.2.1  Bremsstrahlung Radiation .............................. 65
   10.3   Photons ................................................... 66
          10.3.1  Half Value Layer ..................................... 66
          10.3.2  The Interaction of Photons with Matter .................... 67
          10.3.3  Build up ........................................... 70
   10.4   Neutrons .................................................. 70
          10.4.1  Sources of Neutrons .................................. 71
          10.4.2  Neutron Energy Groups ............................... 71
          10.4.3  Elastic Scattering (n, n) .............................. 72
          10.4.4  Inelastic Scattering (n, n') ............................ 72
          10.4.5  Radiative Capture Reactions (n, $\lambda$) ..................... 73
          10.4.6  Moderation ......................................... 73
          10.4.7  General Principles of Neutron Shielding .................. 73
   10.5   Chapter Summary ........................................... 73
   10.6   End of Chapter Questions ..................................... 74

**11    Operational Radiation Protection** ............................... 75
   11.1   Principles of Protection ICRP ................................. 76
   11.2   Radiation Workers .......................................... 77
   11.3   Areas Containing a Radiation Hazard ........................... 78
   11.4   Summary .................................................. 78
   11.5   Exercises .................................................. 79

**12  Nuclear Emergencies** .............................................   81
     12.1  Nuclear Reactor Overview  .....................................   81
     12.2  How to Respond  ..............................................   83
     12.3  Nuclear Emergencies in Perspective  ...........................   84
     12.4  A Chronological Summary of Major Nuclear Emergencies  ..........   85
     12.5  Summary  .....................................................   86
     12.6  Exercises  ....................................................   86

**Solutions to End of Chapter Exercises** .....................................   87

**Key Organisations** .......................................................   95

**Terminology**  ...........................................................   99

**Scientific Notation** ......................................................   101

**Half-Lives of Common Or Useful Radioisotopes** .............................   103

**Comparative Whole-Body Effective Doses**  ...............................   105

**Further Reading**  ........................................................   107

**Bibliography** ...........................................................   109

# About the Author

**Jordan Wain** B.Sc. (Hons) PGCHE, M.Sc. M.Phil. (Cantab) CRadP MSRP holds three degrees in Chemistry, The Physics and Technology of Nuclear Reactors, and Nuclear Energy. He is a Chartered Radiation Protection professional and a member of the Society of Radiological Protection. Presently, he is a Senior Lecturer in Nuclear Emergency Response with a research focus in Radiation Protection.

# Acronyms

| | |
|---|---|
| ADS | Approved Dosimetry Service |
| ALARA | As Low As Reasonably Achievable |
| ALARP | As Low As Reasonably Practicable |
| ARS | Acute Radiation Syndrome |
| BEIR | Biological Effects of Ionizing Radiation |
| BSO | Basic Safety Objective |
| CIDI | Central Index of Dose Information |
| COMARE | Committee on Medical Aspects of Radiation in the Environment |
| CoRWM | Committee on Radioactive Waste Management |
| CT | Computed Tomography |
| DDREF | Dose and Doserate Effectiveness Factor |
| DEPZ | Detailed Emergency Planning Zone |
| DNA | Deoxyribonucleic Acid |
| DNSR | Defence Nuclear Safety Regulator |
| EA | Environment Agency |
| EPR16 | Environmental Permitting Regulations 2016 |
| HASWA74 | Health and Safety at Work Act 1974 |
| HAW | Higher Activity Radioactive Waste |
| HECA | Hazard Evaluation and Consequence Assessment |
| HLW | High Level Waste |
| IAEA | International Atomic Energy Agency |
| ICRP | International Commission on Radiological Protection |
| ICRU | International Commission of Radiation Units and Measurements |
| ILW | Intermediate Level Waste |
| IRR17 | Ionising Radiations Regulations 2017 |
| LET | Linear Energy Transfer |
| LLW | Low Level Waste |
| LLWR | Low Level Waste Repository |

| | |
|---|---|
| LNT | Linear Non-Threshold |
| LSS | Life Span Study |
| NORM | Naturally Occurring Radioactive Material |
| NRPB | National Radiological Protection Board |
| NRRW | National Registry of Radiation Workers |
| ONR | Office of Nuclear Regulation |
| OPZ | Outline Planning Zone |
| PHE | Public Health England |
| RBE | Radiobiological Effectiveness |
| REPPIR19 | Radiation (Emergency Preparedness and Public Information) Regulations 2019 |
| RSA93 | Radioactive Substances Act 1993 |
| SAPs | Safety Assessment Principles |
| SEPA | Scottish Environmental Protection Agency |
| SFAIRP | So Far As Is Reasonably Practicable |
| UN | United Nations |
| UNSCEAR | United Nations Scientific Committee on the Effects of Atomic Radiation |
| VLLW | Very Low Level Waste |
| VSLW | Very Short Lived Waste |

# Introductory Nuclear Physics

Radioactive species are intrinsically unstable atoms; ultimately, they have too much energy and, thus, 'decay' by emitting some of this energy, transforming into more stable atoms. The emitted energy will either be a particle (alpha, beta, neutron) or a photon, which is a packet of electromagnetic energy (X-ray, gamma). These radioactive species are often unstable because of their ratio of sub-atomic particles: neutrons, protons, and electrons. Atoms with the same number of protons but different numbers of neutrons are termed *isotopes*. Different isotopes of the same element will exhibit different nuclear properties. For instance, U-238 has a half-life of around 4.5 billion years, whereas U-235 has a half-life of around 700 million years and is more useful as a fuel in nuclear reactors. As these isotopes are the same element, uranium, they will undergo identical chemical reactions, and therefore, the most common method of separating them relies on their different masses.

Generally, radioactive instability becomes more common as atoms get more protons; the atoms with more protons than uranium are termed *transuranics* and are all radioactive.

## 1.1    Types of Ionising Radiation

A neutral atom contains the same number of negatively charged electrons surrounding the nucleus as it does positively charged protons within the nucleus. When a charged particle approaches the atom, it will interact with the orbital electrons, either repelling or attracting them. If the charged particle has enough energy, it may remove electrons from the atom, creating an electron–ion pair in a process known as ionisation. In air, an average of 34 eV of energy is required to remove an electron. Ionising radiation has

© The Author(s), under exclusive license to Springer Nature Switzerland AG 2025
J. Wain, *Ionising Radiation Protection*, Synthesis Lectures on Engineering, Science, and Technology, https://doi.org/10.1007/978-3-031-65525-8_1

sufficient energy to cause ionisation in the medium through which it passes. However, this is a fluid definition, as different materials will require different amounts of energy to undergo ionisation. The types of ionising radiation are:

*Alpha (α) radiation,* an alpha particle is composed of a helium nuclei, with two protons and two neutrons, making it charged and massive by nuclear standards.
*Beta (β) radiation,* a beta particle is a high-speed electron that has been ejected from a radioactive nuclei, being charged but much lighter than alpha particles.
*Gamma (γ) radiation* comprises of high-energy photons emitted by excited nuclei, uncharged but capable of interacting with electrons.
*X-radiation (X-rays)* is composed of high-energy photons identical to those of gamma radiation, produced when atomic electrons change their energy shells.
*Neutron radiation* involves neutrons emitted from radioactive nuclei or produced during nuclear reactions. Neutrons can have a broad spectrum of energies.

All types of radiation possess energy. Classical mechanics shows that a particle's energy is related to its mass (m) and velocity (v) (kinetic energy = ½ $mv^2$). This means that a heavy, fast-moving particle possesses more energy than a light, slow-moving particle. However, classical mechanics becomes less applicable as a particle's size decreases and begins to move faster, in these cases quantum mechanics becomes more relevant.

Gamma and X-ray radiation have no mass and exist as packets of electromagnetic energy termed photons. The energy of electromagnetic radiation is related to its frequency by:

$$E = hf \qquad (1.1)$$

where E is the energy in joules (J), h is Planck's constant ($6.626 \times 10^{-34}$ $Js^{-1}$), and f is the frequency in Hertz (Hz) ($s^{-1}$).

The energy of electromagnetic radiation is related to its wavelength by:

$$E = hc/\lambda$$

where c is the speed of light ($2.998 \times 10^8$ $ms^{-1}$) and $\lambda$ is the wavelength (m).

Accordingly, we can note that electromagnetic radiation with higher energy will have a smaller wavelength. The wavelength of visible light is $3.8 \times 10^{-6}$ to $7 \times 10^{-6}$ m, and the wavelength of gamma radiation produced from the decay of Cs-137 is $1.87 \times 10^{-12}$ m.

## 1.2      Radioactive Decay

Predicting when an individual radioactive atom will decay is impossible; however, we can determine the average lifetime of a group of atoms. We use the terminology of the *half-life*: the time it takes for half of the atoms to decay. The half-life can vary enormously, from billions of years to fractions of a second. Because each half-live relates to halving the number of radioactive atoms the decay follows an exponential curve. An example of this can be seen for Cs-137, which has a half-life of 30 years.

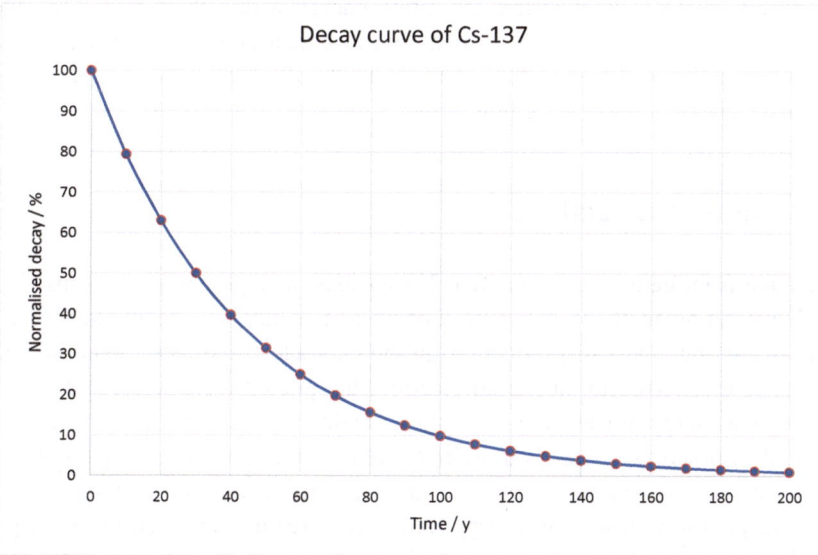

Mathematically, radioactive decay is the change in the number of atoms over a specific time.

$$dN/dt = -\lambda N$$

where
   N is the number of atoms
   t is time, s
   $\lambda$ is the decay constant, $s^{-1}$
The negative sign shows that the number of atoms will decrease over time as they radioactively decay.
   If $N_0$ is the initial number of atoms when t = 0, we can integrate the equation to find:

$$\log(N/N_0) = -\lambda t$$

Thus,

$$N_t = N_0 e^{-\lambda t}$$

Usefully, the number of atoms is proportional to the activity, A, thus:

$$A_t = A_0 e^{-\lambda t}$$

$$\lambda = \ln(2)/t_{1/2}$$

where $t_{1/2}$ is the half-life.

The dose is a result of radiation depositing energy within the tissue. This energy is deposited when subatomic particles and photons lose their energy within that tissue. Consequently, fewer decays result in less dose. Thus, the dose rate (dose per unit time) will also decrease exponentially, according to the radioisotopes' half-life.

## 1.3   Photon Interactions

The radiation from gamma radiation and X-rays exist as high-energy photons. The three essential interaction methods between photons and matter are the photoelectric effect, Compton scattering, and pair production. In the photoelectric effect, the entire energy of the photon is transferred to an orbital electron, the photon disappears, and the electron gains sufficient energy to leave the atom, thus creating an ion. In Compton scattering, part of the incident photon's energy is transferred to an orbital electron; the photon continues with a lower energy and the electron is liberated from the atom. In pair production, the incident photon transfers its energy into mass, creating an electron and a positron. The positron will subsequently interact with another electron and annihilate, producing two 511 keV photons emanating at a 180-degree angle away from one another. For pair production to occur, the incident photon must have an energy exceeding the rest mass of two electrons, 1.022 MeV.

Frequently, more than one type of photon interaction occurs in a material that is being irradiated. The dominant photon interaction is determined by the energy of the incident radiation and the material's atomic number. This may be seen in Fig. 1.1.

## 1.4   Summary

- Ionisation is the process of removing an electron from an atom, creating a free electron and a positively charged ion.
- The types of ionising radiation are alpha, beta, gamma, X-rays, and neutron.
- Different isotopes will exhibit different nuclear properties but identical chemical properties.

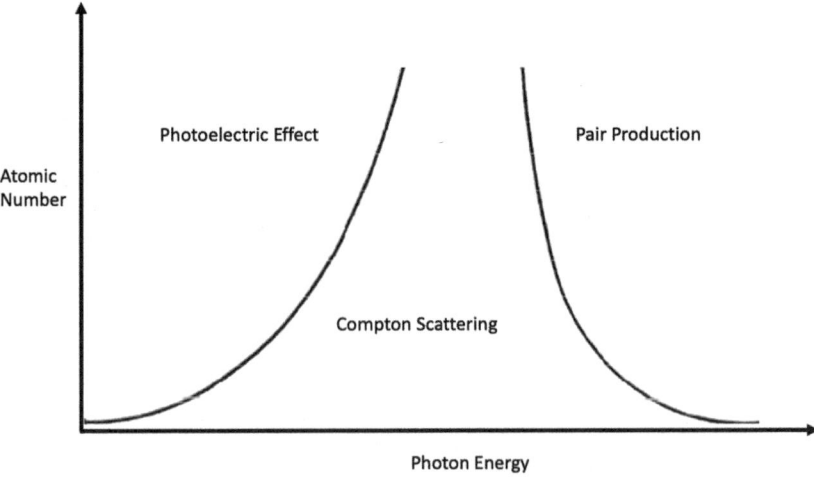

**Fig. 1.1** The dominant photon interaction as a function of energy and atomic number

- Radioactive decay is described in terms of half-lives. One half-life is the amount of time it takes for half of the atoms to decay.
- Gamma radiation and X-rays are high-energy photons. They interact with matter through the photoelectric effect, Compton scattering, and pair production.

## 1.5    Exercises

1. What principle is commonly used to separate U-238 from U-235?
2. What is the theoretical minimum energy required for pair production?
3. What is the decay constant for Cs-137 with a half-life of 30 years?
4. How long will it take for 95% of a sample of Cs-137 to decay?
5. What is the difference between X-rays and gamma radiation?

# Key Units

## 2.1 Activity, Bq

The activity of a source is measured in *Becquerels* (Bq), which is defined as *the number of nuclear decays per second*. Accordingly, the activity will be related to the half-life of the radioisotope. A small number of atoms of a short-lived radioisotope can have a similar activity to a large number of atoms of a longer-lived radioisotope. Using the following equation, we can calculate the activity:

$$A = \lambda N \qquad (2.1)$$

where:

A—activity, Bq

$\lambda$—decay constant, $s^{-1}$

N—number of atoms.

We can calculate the number of atoms in a certain mass of a particular species using the following equation:

$$N = m/MN_A \qquad (2.2)$$

where:

m—mass

M—the molar mass

$N_A$—Avagadro's constant, $6.022 \times 10^{23}$ atoms $mole^{-1}$

J. Wain, *Ionising Radiation Protection*, Synthesis Lectures on Engineering, Science, and Technology, https://doi.org/10.1007/978-3-031-65525-8_2

This enables us to calculate the *specific activity, $A_{sp}$, the activity per unit mass* often expressed as Bq g$^{-1}$.

Some countries use Curies to express activity, one Curie $= 3.7 \times 10^{10}$ Bq.

## 2.2    Source Strength, Q

Some radioisotopes will produce more than one photon or particle per disintegration. A classic example of this is $^{60}$Co, which produces two $\gamma$ photons, of 1.17 MeV and 1.33 MeV, for each disintegration. Thus, $^{60}$Co has a Q $= 2 \times$ activity.

## 2.3    Linear Energy Transfer (LET)

LET measures the amount of energy deposited by a particle or photon as it travels. A high LET means a large amount of energy is deposited over a short distance. Alpha radiation has a high LET, so it is very easy to shield. The concept of LET is essential when considering how much biological damage different types and energies of ionising radiation cause and relates to *radiobiological effectiveness*.

## 2.4    Absorbed Dose, D$_{abs}$

Absorbed dose is the amount of energy deposited per unit mass and is measured in Grays (Gy) or Joules/Kilogram (J/Kg). We use D$_{abs}$ to determine the level of harmful tissue reactions.

## 2.5    Equivalent Dose, H$_T$

Equivalent dose accounts for how the differences in the type of radiation will affect the likelihood of risk based effects (previously known as *stochastic* effects), like cancers. We use dimensionless radiation weighting factors, W$_R$, to account for this. Equivalent dose is measured in Sieverts (Sv) and since the W$_R$ are dimensionless, Sv are also measured in J/Kg (Table 2.1).

$$H_T = D_{abs} \cdot W_R$$

**Table 2.1**  Radiation weighting factors[a]

| Photons (γ, x-rays) | 1 |
|---|---|
| Electrons (β) | 1 |
| Alpha particules (α) | 20 |
| Neutrons | 2.5–20 (dependent upon neutron energy, peaking at 1 MeV) |

## 2.6  Effective Dose, E

Effective dose accounts for how different tissues within the human body have different propensities for developing cancer. Typically, tissues that contain rapidly dividing cells are more radiosensitive; thus, bone marrow, responsible for producing red blood cells, is more radiosensitive than bone surfaces, which grow very slowly. We use tissue weighting factors, $W_T$, to account for this. E gives us the most precise indication of the likelihood of developing risk based effects, like cancers, resulting from exposure. Effective dose is also measured in Sieverts (Sv) and since the $W_T$ do not have any units, Sv are also measured in J/Kg (Table 2.2).

**Table 2.2**  Tissue weighting factors[a]

| Bone surface | 0.01 |
|---|---|
| Skin | 0.01 |
| Bladder | 0.04 |
| Breast | 0.12 |
| Liver | 0.04 |
| Oesophagus | 0.04 |
| Thyroid | 0.04 |
| Bone marrow | 0.12 |
| Colon | 0.12 |
| Lung | 0.12 |
| Stomach | 0.12 |
| Gonads | 0.08 |
| Brain | 0.01 |
| Salivary Glands | 0.01 |
| Remainder | 0.12 |
| Total | 1.0 |

## 2.7     Chapter Summary

- Activity is measured in Bq, the number of nuclear decays per second, or Curries. One Currie is $3.7 \times 10^{10}$ Bq.
- LET defines how much energy is deposited over a certain distance, which indicates the concentration of energy deposition.
- Absorbed dose is measured in Gy and is the amount of energy deposited within a certain mass. It is used to indicate harmful tissue reactions.
- Equivalent dose is measured in Sv and is the absorbed dose multiplied by a radiation weighting factor.
- Effective dose is also measured in Sv and is the equivalent dose multiplied by a tissue weighting factor. The effective dose is used to indicate the increased probability of developing cancer.

## 2.8     End of Chapter Exercises

1. What mass of $^{238}$U would be required for a 1 GBq source?
2. What mass of $^{60}$Co would be required for a 1 GBq source?
3. What is the absorbed dose from a 14 J exposure to a 70 kg adult?
4. What will be the whole-body effective dose if an individual receives a 3 Gy absorbed dose to their entire body from a gamma-emitting radioisotope?
5. What will be the whole-body effective dose if an individual receives a 30 mGy exposure to their lungs from an alpha-emitting radioisotope?

# Cellular and Whole-Body Responses to Radiation

**3**

Most human cells will contain a nucleus, cytoplasm, and cell membrane. The nucleus of the cell contains the cell's chromosomes, and these chromosomes contain deoxyribonucleic acid (DNA) (Fig. 3.1). DNA is a sequence of four chemical bases: adenine, guanine, cytosine and thymine. These bases pair up with one another to form base pairs. Each base is also attached to a sugar molecule and a phosphate molecule; together, these are known as a nucleotide. Nucleotides exist as long stands that spiral around each other in a double helix. The DNA contains the information required for the cell to replicate; consequently, the nucleus is the most radiosensitive part of the cell. Most human cells replicate via the process of mitosis.

Mitosis entails splitting a singular cell into two genetically identical daughter cells. This intricate process encompasses the duplication of genetic information within the cell followed by its partitioning into the newly formed cells essential for bodily growth, repair, and the renewal of aged or impaired cells.

During the stages of mitosis cells are more radiosensitive: it is more likely that energy deposited into the cell from ionising radiation will damage the DNA. Consequently, organs that comprise rapidly dividing cells are more radiosensitive than those that contain very slowly dividing cells.

As children are still developing, they have a faster overall rate of cell replication. This faster replication is one of the reasons children are more radiosensitive than adults. The secondary key reason is life expectancy: children have longer to live, and so there is longer for small cell mutations to develop into adverse health effects, like cancerous solid tumours.

Ionising radiation can damage a cell either directly or indirectly. Direct damage refers to the influx of energy in a particular region damaging that region, for instance the direct

© The Author(s), under exclusive license to Springer Nature Switzerland AG 2025    11
J. Wain, *Ionising Radiation Protection*, Synthesis Lectures on Engineering, Science, and
Technology, https://doi.org/10.1007/978-3-031-65525-8_3

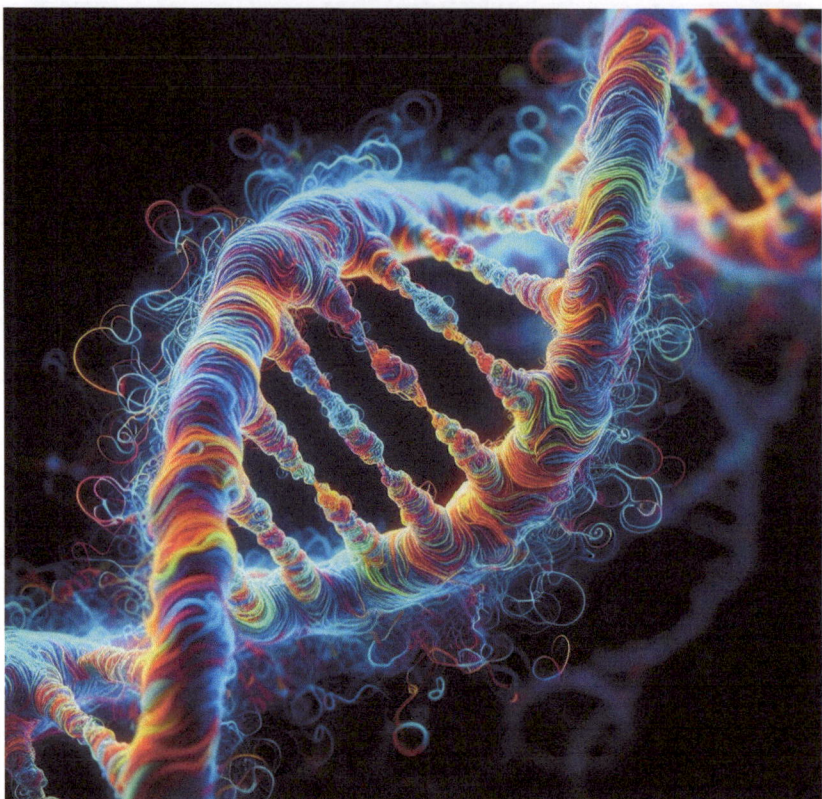

**Fig. 3.1** DNA double helix

irradiation of a DNA molecule; however, ionising radiation can also ionise molecules within the cell, creating secondary molecules, which are much more chemically reactive. As most of the cell is water, indirect damage is considerably more likely. An example of indirect damage is radiation ionising a water molecule, creating hydrogen peroxide and subsequently forming hydroxide radicals:

$$2H_2O + O_2 + \text{ionising radiation} \rightarrow 2H_2O_2 \rightarrow 4HO^{\cdot}$$

When energy from ionising radiation damages a cell, there are ultimately four primary possible outcomes:

- The cell will recover
- The cell will die
- Damage to genetic material results in loss of the ability to replicate
- Damage to genetic material results in cell mutations that may result in cancer.

## 3.1    Cell Recovery

Background radiation is primarily naturally occurring radiation emanating from many sources, including radioactive decay of uranium isotopes in common minerals, radiation from space, and radiation exposure from radioactive isotopes in our food and water. This radiation affects all lifeforms on earth and has done so since the dawn of time. Accordingly, every lifeform that has ever existed has evolved mechanisms to mitigate against damage caused by ionising radiation. In humans, there are several mechanisms through which DNA can repair itself after radiation damage; these include.

*Direct Repair*: Some types of DNA damage can be repaired directly by specialised enzymes.

*Base Excision Repair (BER)* is a repair mechanism for minor, non-helix-distorting damage in DNA, such as impaired or mismatched bases. Specialised enzymes recognise and remove the damaged base, and DNA polymerase fills the gap with the correct base.

*Nucleotide Excision Repair (NER):* involves the removal of a segment of the damaged DNA strand followed by resynthesis using the intact complementary strand as a template. It is used to repair more significant damage to the DNA helix.

*Mismatch Repair (MMR)* corrects errors during DNA replication and recombination. It identifies and removes mispaired bases and restores them by re-synthesising the DNA strand.

*Double-Strand Break Repair:* Double-strand breaks are the hardest to repair consistently as the cell no longer has a template to create the repair. There are several repair pathways in the event of a double-strand break, including Non-Homologous End Joining (NHEJ) and Homologous Recombination (HR). NHEJ directly re-joins broken ends, often with minor errors, while HR uses an undamaged DNA strand as a template to repair the break with high fidelity.

DNA repair mechanisms are not always perfect, potentially leading to mutations and, in some cases, cancer. The effectiveness of DNA repair will be influenced by factors such as the type and extent of damage, individual genetics, and the availability of necessary repair enzymes and proteins.

## 3.2    Cell Death

Widespread cell death will result in harmful tissue reactions. Harmful tissue reactions are associated with a threshold value as it takes a certain amount of energy to begin to kill cells in a sufficient quantity for the effects to be noticeable. Beyond this threshold, the severity of harm increases with the radiation dose because a more significant portion of the cells within the tissue or organ are affected. As there are only a finite number of cells to kill, at still higher doses the increased severity of an effect with additional dose will decrease, simply because there are fewer and fewer remaining healthy cells (Fig. 3.2).

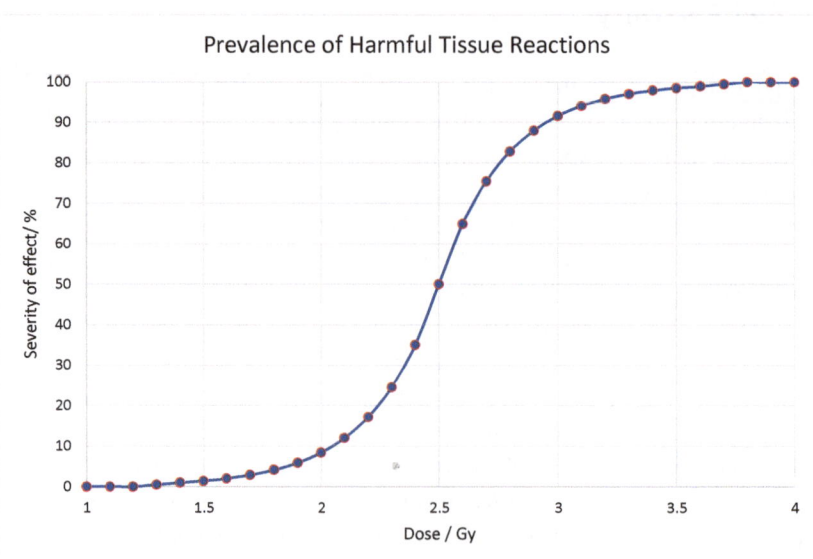

**Fig. 3.2** Prevalence of harmful tissue reactions with dose

Death may result directly from loss of function provided by the damaged tissue or from secondary infections. For example, if an acute whole-body exposure sufficiently damages bone marrow, the result will be aplastic anaemia, a severe hematologic condition where the body cannot create enough blood cells. These blood cells include red blood cells—for transporting oxygen around the body, white blood cells—for fighting infections, and platelets—for preventing and stopping bleeding. The leading cause of death from the loss of these functions is generally dictated by the level of medical care available. In a localised high-dose exposure, for instance, if an individual inadvertently picks up a high-activity radiography source, the dose received will be exponentially proportionally to the distance from the source. Thus, the tissues of the fingers and hand will receive the most significant dose. If this is high enough to kill a sufficient number of cells, dead cells will begin to decompose and may facilitate the entry of bacteria into the bloodstream. Once in the bloodstream, bacteria can be transported to other parts of the body. When the body detects the presence of bacteria in the bloodstream, it can trigger a powerful immune response known as sepsis. Symptoms of sepsis include fever, increased heart rate, rapid breathing, low blood pressure, and altered mental state. Sepsis is a life-threatening condition; thus, a relatively trivial initial injury that may necessitate the amputation of a few fingers can be fatal if left untreated.

Examples of harmful tissue reactions also include radiation burns (acute radiation dermatitis), hair loss (radiation-induced alopecia), damage to the gastrointestinal tract, damage to nerve cells, and cataracts.

Most harmful tissue reactions occur relatively quickly following a sufficiently high radiation exposure. They may present in the days or weeks following the incident, except radiation-induced cataracts, which generally require years, or even decades, to form.

Threshold Doses

| System | Threshold whole body dose (Gy) |
| --- | --- |
| Bone marrow (blood) | 0.5 |
| Skin | 3 |
| Gastrointestinal tract | 5 |
| Central nervous system | 10 |

## 3.3   Loss of the Ability to Replicate

If the cells are damaged to the extent that they can no longer divide but have not died, the exposed individual will have a latency period before any effects are experienced. Different types of cells will have different lifespans, so the exposed individual will effectively run out of various kinds of cells at different times. Typical cell lifespans are:

| White blood cell | 1–3 days |
| --- | --- |
| Gastro-intestinal tract lining | ~15 days |
| Red blood cells | ~100 days |
| Bone cells | ~7 years |
| Brain cells | Lifetime of individual |

## 3.4   Cancer Formation

In our context, it is risk based effects that result in cancers. For radiation damage to occur, the deposited energy must damage the cell's DNA, the cell must not repair this damage, the damage must result in the cell losing its self-regulating ability to control its divisions, and these divisions must continue until a cancerous mass is present. The reader will have identified a lot of variables in this process, making it impossible to accurately predict if an individual damaged cell will become cancerous. Instead, we assign a likelihood to the effect. The International Committee for Radiological Protection (ICRP) lists this likelihood as around 5%/Sv, meaning if someone were to receive a dose of 1 Sv over a reasonably prolonged period, known as a chronic exposure, their risk of developing cancer throughout the rest of their life would be heightened by 5%. If an individual has a baseline risk of developing cancer of 50%, their total risk of developing cancer would increase to 55% as a result of the exposure. However, if this dose was received over a short time at a high dose rate, known as an acute exposure, the ICRP risk coefficient is doubled to about 10%/Sv.

*Linear Non-threshold Model*
The ICRP radiation risk figure of 5%/Sv can be extrapolated downwards, suggesting that even very low doses may increase the likelihood of cancer development; however, there is limited evidence to support this low-dose correlation.

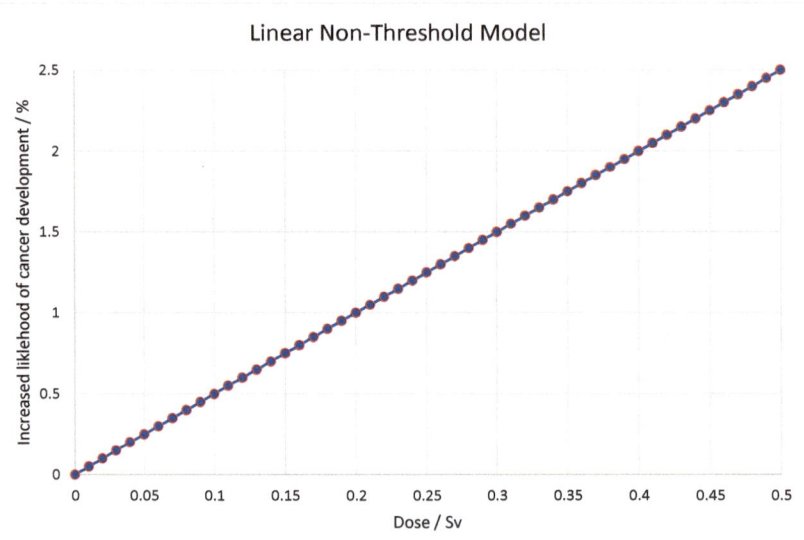

## 3.5    Hereditary Effects

Hereditary effects affect the offspring of the exposed individual. While hereditary effects from radiation exposure have been observed in some animal studies, they have never been observed in humans.

## 3.6    Summary

- DNA is contained within a cell's nucleus.
- DNA can be damaged either directly or indirectly.
- If a cell is irradiated, the result will be cell recovery, cell death, the cell being unable to replicate, or the cell undergoing genetic mutations.

- Harmful tissue reactions will only occur above a threshold value.
- The ICRP assigns a value of about 5%/Sv to describe the increased risk of developing cancer.

## 3.7    Exercises

1. What is the most radiosensitive part of the cell?
2. Describe indirect radiation damage.
3. Why does the prevalence of harmful tissue reactions plateau at higher doses?
4. What is the lifespan of a white blood cell?
5. What will be the increased risk of developing cancer after receiving a chronic dose of 20 mSv?

# Epidemiological Studies and Radiation Risk

4

Epidemiological studies investigate the distribution and determinants of health-related events, conditions, or behaviours within populations. They can be observational or experimental and aim to answer questions about the causes of diseases, risk factors, prevalence, incidence, and the effectiveness of interventions. These studies are widely used to identify and understand health and disease patterns and develop strategies for disease prevention and control.

There are different types of epidemiological studies, but a generalised methodology is to select a specific group exposed to a particular risk factor and then compare this group to an analogous group not exposed to a specific risk factor. Any change in the prevalence of diseases in your initial group may then be correlated to the risk factor. Epidemiological studies often highlight correlations, not necessarily causations.

In the context of cancer induction, the astute reader will understand it is difficult to find the perfect control group: a significant number of individuals with identical diets, exercise regimes, genetic predispositions, smoking preferences, etc.

The total number of epidemiological studies in the field of radiation protection is too numerous to describe fully; however, a summary of the key studies is given below.

## 4.1 Summary of Critical Epidemiological Studies

1. Hiroshima and Nagasaki Studies (Life Span Study): These long-term studies by the Radiation Effects Research Foundation (RERF) have followed survivors of the atomic bombings in Hiroshima and Nagasaki, Japan since 1950 to assess the health effects of acute radiation exposure. Over 120,000 survivors have been included in the studies.

The Life Span Study concludes that individuals exposed to higher radiation doses were more likely to develop various cancers, including leukaemia and solid tumours. The studies demonstrated a clear dose–response relationship for large doses of ionising radiation received over a very short time.[2]

2. Survivor Studies: Various studies have investigated the health effects of the Chornobyl nuclear disaster in 1986, focusing on individuals exposed to radioactive fallout and their risk of cancer and other diseases. Tens of thousands of participants have been involved in various studies. Collectively, these studies conclude that exposed individuals exhibit elevated levels of thyroid cancer, leukaemia, and solid tumours, particularly among children exposed to radioactive iodine isotopes. These studies emphasise the importance of thyroid screening in the aftermath of any nuclear emergency that involves an operational or recently operational nuclear reactor. Iodine-131 has a half-life of around eight days, so it is not a concern if a reactor has not been operational recently.

3. Nuclear Workers Studies (International Nuclear Workers Study—INWORKS): INWORKS is one of the most significant collaborative efforts, examining the health of nuclear industry workers exposed to chronic low doses of radiation from multiple countries. Over 600,000 nuclear workers have participated in the study to date. The study concludes that chronic exposure to low levels of ionising radiation among nuclear workers was correlated with a slightly increased risk of certain cancers, such as leukaemia and lung cancer.

4. Fukushima Health Management Survey: Initiated after the Fukushima Daiichi nuclear disaster in 2011, this study monitors the health of residents and workers exposed to radiation, with a focus on thyroid cancer and other potential health risks. Hundreds of thousands of residents and workers have been included in the study. The study concludes that the overall risk of health effects from exposure to ionising radiation during the Fukushima disaster is low; however, there are significant psychological and social impacts that have been linked to evacuation and relocation.

5. Radon and Miners Studies: These studies have examined the health of underground miners exposed to radon gas, focusing on the risk of lung cancer. Tens of thousands of miners have been included in the studies. They have established a clear link between high levels of radon gas exposure and an increased risk of lung cancer. As a result, regulatory measures have been implemented to reduce radon exposure.

6. Medical Radiation Studies: Various studies have assessed the health risks associated with diagnostic and therapeutic medical radiation, such as CT scans and radiotherapy for cancer treatment. For radiotherapy to be effective in cancer treatments, the radiation levels must be high enough to kill cancerous cells; consequently, the doses received can be very high, in the region of 50 Gy, delivered in separate sessions over months. The exact details will vary with individual patient treatment plans. These studies also link high levels of radiation exposure at high dose rates with cancer induction and highlight the importance of dose optimisation.

7. Radiation Exposure in Airline Crew: One contributor of background radiation is cosmic radiation. This radiation is attenuated by the atmosphere; the higher you are, the greater the dose rate from cosmic radiation. Airline crews spend significantly longer at altitude than any other large group, and several studies involving thousands of aircrew have been conducted; however, these studies have not found strong evidence of significant health risks to airline crew as a result of their exposure to cosmic radiation.
8. Low-Dose Radiation Research: Several studies have attempted to investigate the health effects of low doses of radiation. The studies typically involve thousands of people and have resulted in diverse conclusions. Some studies have indicated that low doses of radiation have a beneficial impact on health. As cells that rapidly divide are generally more radiosensitive, and cancerous mutations result in cells that rapidly divide, one hypothesis is that low doses of radiation preferentially kill cancerous cells before they develop into cancerous growths.

## 4.2    Difficulties with Epidemiological Studies

Unfortunately, epidemiological studies investigating dose–response relationships are intrinsically challenging. The ICRP Publication 103 lists the cancer detriment of radiation exposure for workers as 4.1%/Sv. If we were to extrapolate this figure to low radiation dose levels, we can see the likelihood of cancer induction becomes very small. For instance, the average background radiation level in the UK is 2.7 mSv/y. This equates to an elevated risk of $(5 \times 2.7 \times 10^{-3} = )$ 0.0135%. Currently, in many countries, the likelihood of developing fatal cancer is around 50%, and we are constantly surrounded by carcinogens: cigarettes, processed foods, air pollution, etc. Thus, the tiny increase expected from background radiation would be completely indiscernible compared to the average fluctuations, the 'noise' in a sample. This may be overcome if the study were to find an enormous group to investigate, and it could compare to another perfectly equivalent group with the same diet, smoking preference, genetic predisposition to cancers, etc. The reader will appreciate that this is practically impossible to fully achieve.

Consequently, the results of most use originate from studies with large sample sizes exposed to high radiation levels that have resulted in statistically significant dose–response relationships. These studies all examine events that have resulted in high levels of radiation exposure received over a short time, for instance, from the detonation of nuclear weapons over populated cities, as is the case with the Life Span Studies. However, almost all exposures within the nuclear industry are chronic: low radiation levels received over a long time. Accordingly, it is necessary to translate the effects from these high-dose and high-dose-rate exposures into the effects anticipated from low doses and low-dose-rate exposures. We need a Dose and Dose Rate Effectiveness Factor (DDREF) for this. The ICRP has assigned a DDREF of 2, meaning that chronic exposures are half as likely to lead to cancer inductance as acute exposures.

## 4.3    Validity of Studies

As carcinogens and thus cancers are so prevalent in today's society, a slight increase in cancers within a group can be caused by a plethora of factors. Additionally, it can be impossible to determine if a tumour is radiogenic retrospectively, and there are often confounding factors. Thus, linking correlation to causation is often not straightforward. For instance, a successful public health campaign to reduce smoking will reduce cancer. Conversely, a successful public health campaign that reduces heart disease is likely to increase cancers, as more individuals will live longer and, therefore, have an increased time to develop cancer. Certain nuclear workers may be at an increased risk of developing skin cancer because of workplace exposures, or they may be getting paid more than average and go on more holidays to sunny destinations, receiving more exposure to UV radiation, which results in an increased risk of developing skin cancer—or both.

A typical reliable epidemiological study will investigate a large population size, compare it to an analogous large control group, and highlight statistically significant increases in a specific response.

## 4.4    Healthy Worker Effect

Many epidemiological studies into all aspects of health have concluded that individuals in the workforce are less likely to suffer from some illnesses. Effectively, this stems from a selection bias. By only including employed individuals in your study, you are also only including individuals of a certain age, who are already healthy enough to work, who are likely to have an elevated income compared to those who are too old, too young, or too ill to be included in the study. This is termed the *healthy worker effect*. This has also been observed in the results of epidemiological studies focused on the effects of radiation and has led the ICRP to propose different cancer detriment figures for adult workers and for the whole population (Table 4.1):

Here, the cancer detriment incorporates the detriment from fatal cancer, non-fatal cancer, and a loss-of-life factor. This loss-of-life factor considers the years of life lost and reflects the age at which the individual develops fatal cancer in comparison to their average life expectancy.

**Table 4.1**  ICRP values for the cancer detriment of workers and the whole population[1]

| Exposed population | Cancer detriment %/Sv |
|---|---|
| Adult workers | 4.1 |
| Whole population | 5.5 |

**Table 4.2**  Dose limits for the UK[3]

| Person | Effective dose (mSv) | Skin, hands, forearms feet and ankles | Lens of the eye |
|---|---|---|---|
| | | Equivalent dose (mSv) | Equivalent dose* (mSv) |
| Employee | 20 | 500 | 20 |
| Any other person | 1 | 50 | 15 |

## 4.5    Dose Limits

The dose limits for the UK (Table 4.2).

In the UK, the dose limits were created after three primary considerations. Firstly, they must be below the threshold for any harmful tissue reactions; secondly, for workers, they must be commensurate with a level of risk from similar industries; thirdly, for members of the public, they must not significantly increase their level of risk. It is considered socially acceptable to expose workers to higher levels of risk as they understand the risk, choose to expose themselves to it, and are reimbursed accordingly. This is not the case with members of the public.

## 4.6    Summary

- Epidemiological studies enable us to estimate the overall detriment of radiation exposures.
- The most useful epidemiological studies are from high-dose and high-dose-rate exposures. We translate the radiation detriment from these to low-dose and low-dose-rate situations using a DDREF of 2.
- An epidemiological study's results will be more scientifically valid if it examines a large population group and reports on statistically significant fluctuations.
- The healthy worker effect results from the average health of individuals in employment being better than the average health of the whole population.
- The whole-body effective dose limit for a radiation worker is 20 mSv per year, and the whole-body effective dose limit for a member of the public is 1 mSv per year.

## 4.7    Exercises

1. List some difficulties with epidemiological studies that focus on low-dose chronic exposure.
2. What were the main conclusions from the Radon and Miners studies?
3. What is the average background dose in the UK?
4. What is the increased cancer detriment to a worker from an exposure of 100 mSv?
5. Why does the lens of the eye have a specific dose limit?

# Internal Radiological Exposure

**5**

The hazard posed by ionising radiation may exist externally to the body or internally. External hazards have the significant benefit of being location-dependent; in many cases, it is possible to simply walk away from them. However, internal exposures will continue until either radioactive decay or biological removal mechanisms have eliminated the radioisotope from the body.

## 5.1 Routes of Entry

Radionuclides can enter the body through inhalation, ingestion, injection, and absorption. In this context, injection refers to a radioisotope entering the body through a break in the skin, whether resulting from a stab wound from a thin, sharp, contaminated piece of metal or an abrasion. Absorption is only significant with a few radioisotopes, which exhibit the necessary chemical characteristics to be absorbed through the skin. The most relevant of these is tritium.

Once inside the body, the radioisotope's behaviour will depend upon its chemical form, physical form, and the biological processes acting upon it. For example, metallic oxides are generally very insoluble; therefore, if plutonium oxide is ingested, it will likely remain in the digestive tract and be excreted. However, if the intake was of metallic plutonium, which is more soluble, a more significant proportion would pass from the digestive tract into the blood and from the blood to various organs and tissues where it will be partially deposited, thus dramatically increasing the time that the radioisotope remains in the body. To describe the time that the radionuclide remains in the body, we use the notion of a *biological half-life*, a concept analogous to a radioactive half-life.

J. Wain, *Ionising Radiation Protection*, Synthesis Lectures on Engineering, Science, and Technology, https://doi.org/10.1007/978-3-031-65525-8_5

Numerous studies determining the biological half-life of non-radioactive substances highlight the impact of an individual's specific biological processes. For example, the biological half-life for coffee is between 1.5 and 9.5 h. This extensive range reflects the many factors acting upon the caffeine molecules. The factors may be inherently related to the individual, for instance, their body fat to muscle ratio, sex, and age, or they can be more short-term: how hydrated that individual is at the specific time of the intake, and how stressed they are. Ultimately, anything that affects the individual's metabolism, or anything that may chemically interact with the radioisotope, will influence its biological half-life. In the field of radiation protection, it is often unknown to whom an exposure will occur. Accordingly, making some assumptions about individual differences and using average values is necessary.

To ascertain the total time a radioactive isotope remains in the body, we must, therefore, take account of both the biological half-life, $t_{bio}$, and the radiological half-life, $t_{rad}$. We can combine these to find the *effective half-life, $t_{eff}$*:

$$1/t_{eff} = 1/t_{bio} + 1/t_{rad} \tag{5.1}$$

Figures 5.1, 5.2, and 5.3 illustrate the differences between these half-lives. These figures have been produced using Eq. (5.2), normalised to 100%, to enable the graphical representation of the remaining percentage of the radionuclide.

$$A_t = A_0 e^{-\lambda t} \tag{5.2}$$

$$\lambda = \ln 2/t_{1/2} \tag{5.3}$$

where

$A_t$    is the remaining activity after time, t, has elapsed
$A_0$    is the initial activity
$\lambda$    is the decay constant
$t_{1/2}$    is the respective half-live

A comparison of the radiological, biological, and effective half-lives may be seen in Fig. 5.1 for the normalised decay of Pu-238 over 200 years. The radioactive half-life is 87.7 years, and the biological half-life is 200 years. Two hundred years is an extended timeframe in comparison to life expectancy. Focusing on the first 50 years, we can see from the effective decay curve in Fig. 5.2 that 56% of the total activity will remain and that the exposed individual will receive a reasonably even dose rate for these 50 years. This will result in an even distribution of doses for the remainder of the individual's life. The reader will remember from the previous chapter on the biological effects of radiation that it is preferable to spread out a dose, as this will enable an increased efficacy of

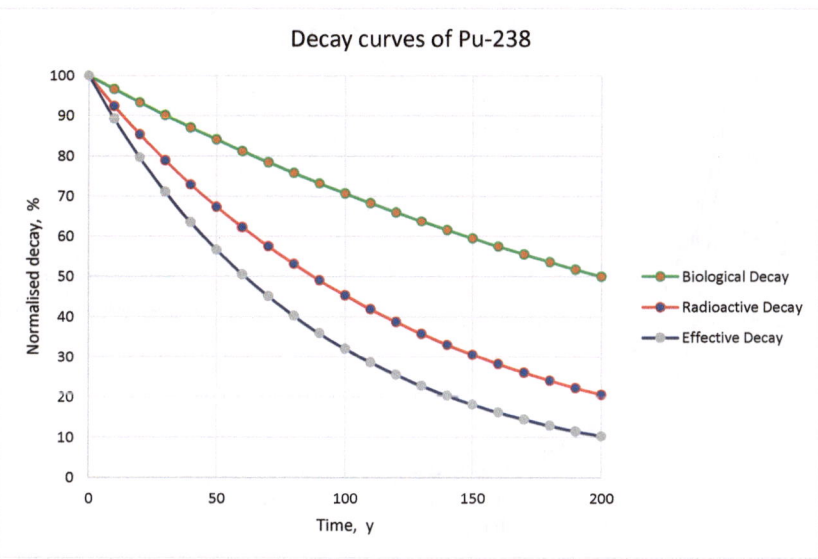

**Fig. 5.1**   Decay curves of Pu-238

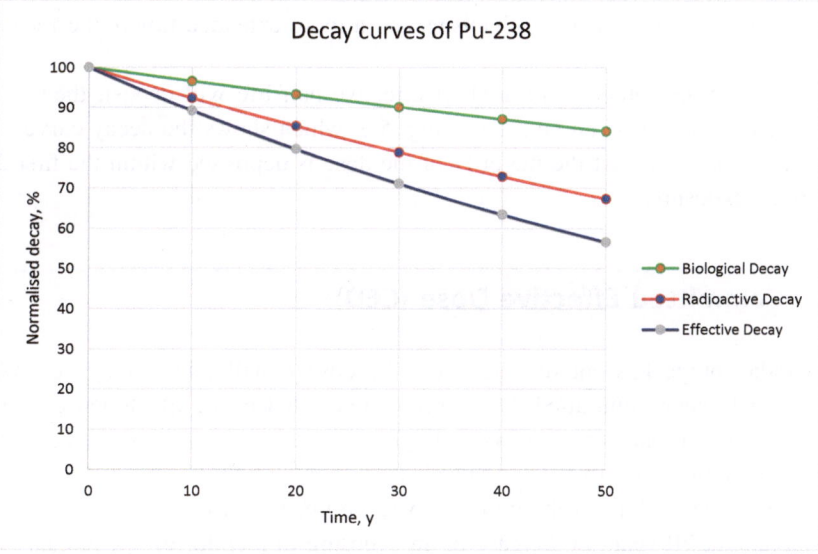

**Fig. 5.2**   Decay curves of Pu-238 to 50 years

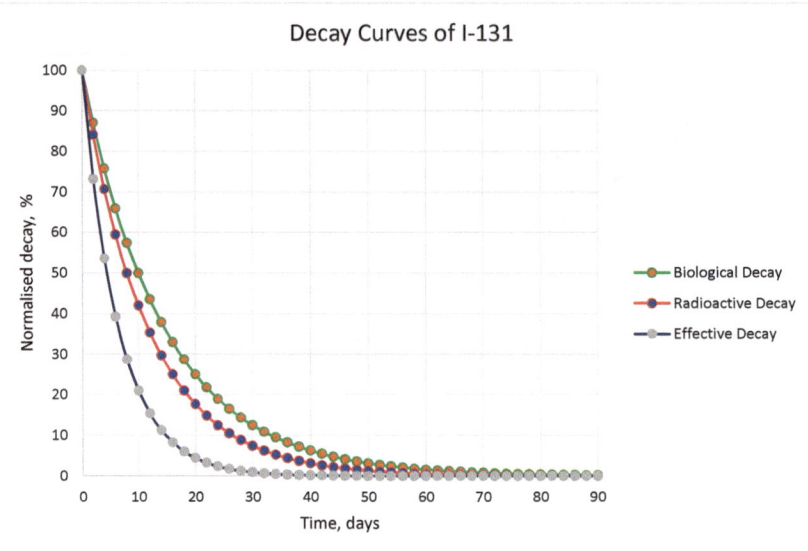

**Fig. 5.3** Decay curves of I-131

enzyme-mediated DNA repair mechanisms and a more extended timeframe for potential medical interventions.

Conversely, radioisotopes with a short effective half-life will deposit their dose in a short space of time. This can be seen in Fig. 5.3, which shows the decay curves for $^{131}$I in days. It can be seen that the majority of the dose is deposited within the first 20 days of the initial exposure.

## 5.2   Committed Effective Dose (CED)

After a radioisotope has entered the body, the dose it will impart may be considered to have already been committed. We are most interested in the likelihood of risk based effects from an internal exposure. Accordingly, we need to understand the effective dose that has been committed to the individual, termed the *committed effective* dose (CED). The CED is calculated as all the dose that will be imparted by the radioisotope over the next 50 years, or 70 years in the case of an exposure to a child. In operational radiation protection, this integrated dose is added to an exposed individual's dose record as though all the dose had been received on the day of intake. This makes the administration of dose records much easier; however, it obscures the reality of the timeframe in which the dose is received, as highlighted in Figs. 5.2 and 5.3.

**Table 5.1**  Effective dose coefficients for inhaled and ingested particulates for workers 4

| Workers | | | Inhalation | | | Ingestion | |
|---|---|---|---|---|---|---|---|
| Nuclide | Half-life | Type | $f_1$ | DPUI 1 μm (Sv Bq$^{-1}$) | DPUI 5 μm (Sv Bq$^{-1}$) | $f_1$ | DPUI (Sv Bq$^{-1}$) |
| I-131 | 8.04 d | F | 1 | 7.60E−09 | 1.10E−08 | 1 | 2.20E−08 |
| Cs-137 | 30.0 y | F | 1 | 4.80E−09 | 6.70E−09 | 1 | 1.30E−08 |
| Pu-238 | 87.74 y | M | 0.0005 | 4.30E−05 | 3.00E−05 | 0.0005 | 2.30E−07 |
| | | S | 0.00001 | 1.50E-05 | 1.10E−05 | 0.00001 | 4.90E−08 |
| Pu-239 | 24,065 y | M | 0.0005 | 4.30E−05 | 3.20E−05 | 0.0005 | 2.50E−07 |
| | | S | 0.00001 | 1.50E−05 | 8.30E−06 | 0.00001 | 5.30E−08 |

To calculate the transport throughout the body of the different chemical forms of various radioisotopes and to then ascertain how much dose is being absorbed by other tissues is a complex task. Fortunately, these calculations have already been performed on most radioisotopes for a *reference person, and the Dose Per Unit Intake (DPUI) has been provided for workers and members of the public, divided into infants, 1-year-olds, 10-year*-olds, and adults. DPUI values enable an expedient conversion from the activity internally deposited to the dose that will be accrued. In Table 5.1, we can also see columns labelled 'Type', which refers to the speed of absorption of the radioisotope in its current chemical form: fast, moderate, or slow; '$f_1$', which refers to the fraction of an element absorbed directly into body fluids; and 1 μm or 5 μm, which refers to the average diameter of the inhaled particle. DPUI values have the units of Sv Bq$^{-1}$, which enables an easy conversion:

$$CED = DPUI \times A \tag{5.4}$$

where

A    is the activity in Bq

Full tabulated values for other radioisotopes can be found in ICRP publications. Comparing the F-type values for adults for caesium-137 and plutonium-239 highlights the different internal risks posed by different radioisotopes. As an example, if an adult had a 10 kBq inhalation intake of F-type, 1 μm particulate, caesium-137, they would receive a CED that is equivalent to around a week of background radiation:

$$4.8 \times 10^{-9} \times 10 \times 10^3 = 48 \ \mu Sv$$

However, if an adult were to have a 10 kBq inhalation intake of S-type, 1 μm particulate, plutonium-239, they would receive a CED that is seven and a half times greater than the dose limit for a radiation worker in the UK:

$$1.5 \times 10^{-5} \times 10 \times 10^3 = 150 \text{ mSv}$$

To further highlight the dangers of some radionuclides, we may calculate the weight of 10 kBq of Pu-239. We can first calculate the total number of Pu-239 atoms present, N, and then calculate how much these atoms will collectively weigh:

$$A = \lambda N \tag{5.5}$$

$$N = A / \left( \ln 2 / t_{1/2} \right)$$
$$= 3.47 \times 10^8 \text{ atoms}$$
$$N = \frac{m}{M} \times N_A \tag{5.6}$$

where

m       is the mass, g
M       is the molar mass
$N_A$    is Avogadro's number, $6.022 \times 10^{23}$ $\text{mol}^{-1}$

As the molar mass is within a fraction of the mass number, we can use the mass number in its place.

$$m = N / N_A \times M$$
$$= 1.38 \times 10^{-13} \text{ g}$$

This is less than the weight of a single particle of household dust.

From Eqs. (5.1) and (5.2), we can see the dose received from a mass of radionuclide will be dependent upon the specific activity and the DPUI, which will dictate the committed effective dose received from this activity. This is further illustrated in Table 5.2, which shows the specific activity and the CED for 1 g of each nuclide.

### 5.2.1    The Importance of Particle Diameter for Inhalation CEDs

The inhalation pathway is relevant for radionuclides present as gases, vapours, aerosols, and solid particulates. An aerosol is an assembly of liquid or solid particles suspended in a gaseous medium. The diameters of airborne particles cover a wide range from ~0.001 to 100 μm. The behaviour of smaller particles is dominated by Brownian motion, whereas larger particles are more affected by gravitational settling. Brownian motion is an erratic, random motion of small particles caused by collision with other small surrounding particles [5]. Whilst the shape of a particle may influence its behaviour, the most critical

**Table 5.2**  Nuclide comparison of specific activities and CED [4]

| Nuclide | Half-life | Specific activity (Bq g$^{-1}$) | DPUI[a] (Sv Bq$^{-1}$) | CED (Sv) |
|---------|-----------|-------------------------------|-----------------------|----------|
| H-3     | 12.35 y   | 3.57E+14                      | 1.80E−15              | 6.43E−01 |
| I-131   | 8.04 d    | 4.59E+15                      | 7.60E−09              | 3.49E+07 |
| Cs-137  | 30.0 y    | 3.22E+12                      | 4.80E−09              | 1.55E+04 |
| U-238   | 4.47E9 y  | 1.24E+04                      | 4.90E−07              | 6.10E−03 |
| Pu-238  | 87.74 y   | 6.34E+11                      | 4.30E−05              | 2.73E+07 |
| Pu-239  | 24,065 y  | 2.30E+09                      | 4.30E−05              | 9.90E+04 |

[a] DPUI values for worker inhalation for 1 μm diameter particulates, F or M type. H-3 values for inhalation of tritium gas

parameter is its diameter. The behaviour of the particle will be correlated to its deposition within the lungs, where deposition is defined as the inhaled quantity minus the amount exhaled. Within the lung, there are three main particulate behavioural mechanisms: Brownian motion, sedimentation, and inertial impaction. Brownian motion may be considered a 'thermodynamic effect', whereas sedimentation and inertial impaction are 'aerodynamic effects'.

The average kinetic energy of these tiny particles can be correlated to their temperature by:

$$E = k_B T \tag{5.7}$$

where

E    is the energy, J
$k_b$   is Boltzmann's Constant, $1.38 \times 10^{-23}$ J K$^{-1}$
T    is the temperature, K

Sedimentation is comparable to gravitational settling, whereby larger, heavier particles are less buoyant and will displace smaller, lighter particles. Inertial impaction relies on momentum. Larger, heavier particles have greater momentum and are thus more easily deposited when the trajectory of the aerosol flow changes, for instance, as inhalation forces air from the trachea into the bronchus.

Brownian motion will dominate the particle behaviour below a diameter size of 0.1 μm and become ever more pronounced as the particle size decreases. Sedimentation and inertial impaction dominate above 1 μm, and effects increase with increasing particle size and density. The result is that smaller diameter particles will travel much deeper into the lungs. Between 0.1 and 1 μm, aerodynamic and thermodynamic effects are present.

In reality, aerosols typically consist of a mix of complex particle shapes. To simplify this, aerosols are defined by a particle size that emulates the behaviour of that aerosol, an equivalent particle diameter.

There are two main biological removal mechanisms from the lungs: coughing up radionuclide-containing mucus and radionuclides passing from the lungs into the blood. Smaller particles can penetrate much deeper into the lungs, dissolve more efficiently, and are more likely to pass into the blood. Larger diameter particles are deposited higher in the respiratory system due to inertial impaction and gravitational settling. Consequently, they are easier to remove with mucus removal mechanisms. Therefore, the most extended biological half-life from inhalation will result from insoluble small-diameter particles.

## 5.3    Mitigating the Internal Radiation Hazard

Avoiding the internal radiation hazard is the best way to mitigate it entirely. Do not work with material that could pose an internal contamination hazard. If this is not possible, use engineered controls to exclude workers from the location of the hazard, use managerial controls to enable workers to exclude themselves when the hazard is present, and finally, use personal protective equipment (PPE) in the form of respirators, over suits, gloves, and leather aprons, as a last line of defence.

However, if these methods fail, and an individual receives an internal exposure, some procedures may reduce the CED. These methods must be tailored to the intake pathway and often to the radioisotope of most concern. They act to reduce the time the radioactivity is in the body and thus reduce the biological half-life. There are many theoretical ways of achieving this, but the most efficacious and experimentally proven are.

*Chelation Therapy*
A chelate is a chemical term referring to a molecule that can bind with a metal atom at two or more places. Specific chelates bond with specific metal atoms. Chelating a metal atom can effectively trap that atom within a larger molecular structure, reducing its mobility and chemical reactiveness. Following the 1987 Brazilian Goiânia incident, Prussian Blue (potassium ferric hexacyanoferrate) was injected into individuals with high levels of internal $^{137}$Cs contamination to chelate the caesium[6].

*Lung Lavage*
Following a severe inhalation intake, it is possible to wash out the radioisotope. This is generally performed using a saline solution to fill each lung in turn while the other lung is ventilated. The solution is drained out of the patient, and the procedure is repeated as required.

*Wound Excision*

Following a contaminated injection wound, it is possible to surgically remove the tissue surrounding the wound site. This will also remove much of the radioactivity that has been transferred into the patient.

These procedures have advantages and disadvantages, and their use must always be justified when weighed against the risk of radioactive contamination. The efficacy of these procedures will depend on many factors, the most important of which is the time from contamination to treatment.

## 5.4   Summary

The key concepts discussed in this chapter are:

- The four routes for radionuclide entry into the body are ingestion, inhalation, injection, and absorption.
- The route of entry, chemical, and physical properties will influence the biological half-life.
- The DPUI and the specific activity result in significant differences in the mass of a radionuclide required to give a particular dose.
- Inhaled small particles will penetrate deep into the lungs. Brownian motion primarily governs their behaviour.
- Inhaled larger particles will deposit higher in the respiratory tract, as their behaviour is dominated by inertial impaction and gravitational settling.

## 5.5   Exercises

1. The radioactive half-life of tritium is 12.35 years, the biological half-life is ten days, what is the effective half-life?
2. How long will it take for 90% of a sample of tritium to decay?
3. How long will it take for 90% of tritium to leave the body after an intake?
4. What is the CED to a worker after ingesting 10 MBq of $^{137}$Cs?
5. What is the mass of $^{238}$Pu, in oxide form, that will result in a CED of 5 Sv if inhaled as 1 $\mu$m particles?

Intuitively, a continuous function would ...

... the ...

In this ...

However, ... these ...

... these ...

## 5.2    Summary

In this chapter ...

* ...
  and ...
* ...
  ...
* ...
  ...
* ...
  ...
* ...
  ...

## 5.3    Exercises

1. ...
2. ...
3. ...
4. ...
5. ...

# External Radiological Exposure

**6**

The dose that originates from radioisotopes outside the body is easier to control. The three main methods are to decrease the time spent in the vicinity of the source, increase the distance from the source, and improve the shielding around the source: time, distance, and shielding. Shielding is covered in more detail in its own chapter.

## 6.1 Time

The reduction of time is the easiest to quantify; if you spend only 15 s at a dose rate of 4 Sv/h, you will receive only $(4/60/60 \times 15 =)$ 16 mSv. This is a linear relationship; if you double this time, you will double your dose.

## 6.2 Distance

Distance can be more complex as it depends on the shape of the source. Fortunately, if the source is not collimated and we are more than three times the size of the face of the source away from it, we can justify treating it as a point source. Collimation is a process that creates a parallel beam of radiation, more similar to a laser than a lightbulb. A point source is an infinitesimally small point that radiates photons/particles uniformly in all directions, 360°. If you could somehow slow time and see the radiation, you'd notice a sphere expanding from the source, inflating much like a perfectly spherical beach ball. The photons/particles emanating from the sphere thus must spread out as they get further from the source. The surface area of a sphere can be calculated as follows:

J. Wain, *Ionising Radiation Protection*, Synthesis Lectures on Engineering, Science, and Technology, https://doi.org/10.1007/978-3-031-65525-8_6

$$A = 4\pi r^2$$

where:

A—the surface area

r—the radius of the sphere

This leads to an inverse square relationship. The $r^2$ term means that as you double the distance from the source, you quarter the dose rate. Mathematically:

$$D_1 \cdot x_1^2 = D_2 \cdot x_2^2$$

where:

$D_1$—initial dose

$X_1$—initial distance

$D_2$—secondary dose

$X_2$—secondary distance

Operationally, this is a quick and relatively accurate method of calculating the dose rate at any distance after you have ascertained the dose rate at a specific distance.

*Example Scenario*

A high-activity source has fallen out of its shielding. You are required to place the source back in its shielding using a 1 m-long pick-up tool. You choose to approach cautiously from a long distance away with a handheld radiation instrument. At 10 m, the instrument indicates a dose rate of 19 µSv/h. What dose will you accrue from returning the source to its shielding?

First, we can calculate the dose rate at 1 m:

**Source**                          **1 m**                                    **10 m**

                                                                                 **19 µSv / h**

We may use the inverse square law:

$D_1$—19 µSv/h

$X_1$—10 m

$D_2$—Unknown

$X_2$—1 m

$$D_2 = D_1 \cdot x_1^2 / x_2^2$$

$D_2 = 19 \times 10^{-6} \times 10^2/1^2$

$D_2 = 0.0019$ Sv/h $= 1.9$ mSv/h

Second, we must calculate the time.

We can make the pessimistic assumption it will take one minute to place the source back in its shielding. Therefore, the total dose received will be 1.9 mSv/60 = 0.032 mSv = 32 μSv.

This is not a perfect calculation; we have neglected the time required to move to the source location, any increased dose from reflections within the room, and any decreased dose from air attenuation, and we have overestimated the time required to return the source to it housing Beyond around 20 m, air attenuation will begin to become more significant and we will need to take this into account.

## 6.3    Summary

- To reduce the dose emanating from external radiation sources, we must decrease the time spent in the vicinity of the source, increase the distance to the source, and improve the shielding around the source.
- The inverse square law is:

$$D_1.x_1^2 = D_2.x_2^2.$$

## 6.4    Exercises

1. How much dose will be received after spending 20 min in a dose rate of 80 mSv/h?
2. How much dose will be received after spending 8 s in a dose rate of 6.4 Sv/h?
3. The dose rate is 4 mSv/h at 2.3 m; what is the dose rate at 9 m?
4. The dose rate is 2 Sv/h at 1.4 m; at what distance will the dose rate be 100 mSv/h?
5. The dose rate is 1.6 Sv/h at 4.2 m; at what distance will you accue a dose of 200 mSv in 15 min?

# Background Radiation

7

Background radiation refers to the ionising radiation constantly present in the environment. Since the dawn of life on Earth every individual, from every species, that has ever existed, has been ever continuously exposed to background radiation. Today, the sources of background radiation may be categorised as primordial, cosmic, and anthropogenic.

## 7.1 Primordial Background Radiation

Primordial background radiation originates from radioisotopes incorporated into the Earth when it formed about 4.5 billion years ago and are still present in measurable quantities. To exist over geological timescales, these radioisotopes all have very long half-lives and consequently have low specific activities. These primordial radioisotopes decay, forming daughter products that are also radioactive. These daughter products will decay, forming their own daughter products, creating decay chains that end with a stable isotope.

The three most important primordial radioisotopes are $^{238}$U, $^{235}$U, and $^{232}$Th, known as the uranium, actinium, and thorium series, respectively. These series go through around 20 daughter products, end in a stable isotope of lead, and contain an isotope of radon. Radon is a chemically stable alpha emitter and is especially important as it exists as a gas and as a gas has dramatically increased mobility. The following table sequentially shows the major isotopes in the uranium decay series:

© The Author(s), under exclusive license to Springer Nature Switzerland AG 2025     39
J. Wain, *Ionising Radiation Protection*, Synthesis Lectures on Engineering, Science, and
Technology, https://doi.org/10.1007/978-3-031-65525-8_7

| Isotope | Half-Life | Decay mode |
|---|---|---|
| Uranium-238 (U-238) | 4.5 billion years | α (alpha) decay |
| Thorium-234 (Th-234) | 24.1 days | β (beta) decay |
| Protactinium-234 m (Pa-234 m) | 1.17 min | β (beta) decay |
| Uranium-234 (U-234) | 245,500 years | α (alpha) decay |
| Thorium-230 (Th-230) | 75,380 years | α (alpha) decay |
| Radium-226 (Ra-226) | 1,600 years | α (alpha) decay |
| Radon-222 (Rn-222) | 3.8 days | α (alpha) decay |
| Polonium-218 (Po-218) | 3.1 min | α (alpha) decay |
| Lead-214 (Pb-214) | 26.8 min | β (beta) decay |
| Bismuth-214 (Bi-214) | 19.9 min | β (beta) decay |
| Lead-210 (Pb-210) | 22.3 years | β (beta) decay |
| Bismuth-210 (Bi-210) | 5.01 days | β (beta) decay |
| Polonium-210 (Po-210) | 138 days | α (alpha) decay |

Uranium is ubiquitous throughout the Earth's crust; it is around 40 times more common than silver. It is present in soil, minerals, and oceans; however, its concentration varies widely. Due to the chemistry of uranium it is incompatible with volcanic magmas. Volcanic magmas that have cooled and solidified are known as igneous rocks. Granite is an intrusive igneous rock, it is formed when magma cools and solidifies underground, sometimes in cracks or fissures of other rocks, and is accordingly rich in uranium. As the uranium within the granite bedrock slowly decays, it will form various daughter products that will also be trapped in the granite until radon gas is formed. The radon can accumulate in fissures within the granite and be forced to the surface. Usually, the radon will be quickly diluted in the lower atmosphere and then decay into one of its solid daughter products; however, if a house is unfortuitously positioned on top of one of these fissures, the radon gas can instead accumulate in the house. The total background radiation in areas with granite bedrock is often significantly higher when compared to regions with alternative geology.

### 7.1.1  Oklo Reactor

Uranium within uranium-containing ores exists in several isotopes. The most common of these are $^{238}$U and $^{235}$U. $^{235}$U is the major energy-producing radioisotope within almost all the world's power-producing nuclear reactors because $^{235}$U is fissile. The term fissionable describes a nuclide capable of undergoing nuclear fission after capturing a neutron of any energy, even with a low probability. The term fissile describes a nuclide that will undergo fission with a high probability after capturing a neutron of low energy. Both $^{238}$U and $^{235}$U are fissionable; however, only $^{235}$U is fissile. With $^{235}$U, the binding energy resulting from the absorption of a neutron is greater than the critical energy required to split the $^{235}$U atom, the process of fission. Resultantly, $^{235}$U can undergo fission after absorbing a low-energy 'thermal' neutron. Low-energy neutrons have a greater probability of interaction with matter, so the fission of $^{235}$U is comparatively easy to achieve. However, neutrons created from fission have a high energy. This creates the need for a moderator, something that can reduce the energy of the neutrons. Water is a common moderator in modern-day reactors. Conversely, $^{238}$U needs energy greater than the binding energy of a neutron, so it will only fission after absorption of fast neutrons, which have a much lower probability of interaction.

The half-life of $^{238}$U is around 4.5 billion years, about the age of Earth, and the half-life of $^{235}$Uis around 700 million years. Today, $^{235}$U comprises 0.72% of all uranium; naturally, the concentration of $^{235}$U used to be higher.

The Oklo reactors were discovered in 1972 in the Oklo region of Gabon, Africa when French physicists noticed anomalies in the isotopic composition of uranium in samples collected from a uranium mine. Investigations revealed that the anomalies resulted from natural nuclear fission reactions in uranium deposits around 1.7 billion years ago.

Investigations concluded that when a uranium-rich mineral deposit was immersed in groundwater, the groundwater acted as a neutron moderator, enabling a fission chain reaction—the resultant energy created heat, which evaporated the groundwater. With no more moderation, the fission reactions stopped. After cooling, the water returned, reinitiating the reaction. The cycles of fission reactions persisted for hundreds of thousands of years, concluding as the diminishing fissile materials, combined with the accumulation of neutron poisons, prevented the sustained continuation of the chain reaction.

## 7.1.2   Naturally Occurring Radioactive Material (NORM)

Taken literally, the term NORM includes all naturally occurring radioactive material found in the environment. However, this material rarely presents a hazard worth mitigating. The term NORM is used instead to refer to naturally occurring radioactive material where the risk of exposure has been increased by human activity.

NORM can be encountered in many industries, especially those that uncover and process natural minerals or deposits, such as the mining and combustion of fossil fuels, the mining and processing of metal ores, the fertiliser industry, and recycling. The hazard is created either from the exposure of workers at the time of extraction, for instance, miners being exposed to greater levels of radon that have accumulated in the mine, or from the concentration in some way of the radioisotopes, for example, in the slag that forms on top of molten metal as it is smelted.

The main issue with NORM is the prodigious volume produced. The most significant single contributor to NORM is coal ash, estimated at 280 million tonnes being created annually globally [7]. This coal ash contains concentrated radioisotopes from the uranium, actinium, and thorium decay series.

## 7.2    Cosmic Background Radiation

Cosmic radiation consists of high-energy particles of an extraterrestrial origin. The closest source is the sun, which emits primarily alpha particles and protons with a variability that depends upon solar activity. Cosmic radiation that originates from beyond our solar system is termed *galactic radiation*. Galactic radiation consists mainly of electrons and protons created in ancient, distant supernovas. As galactic radiation is charged, it is regulated by the magnetic field of the sun and the heliosphere and thus is less variable than solar cosmic radiation.

These primary particles can enter the Earth's atmosphere with relativistic energies and collide with atmospheric species, creating a particle cascade and cosmic rays. These cosmic rays are attenuated by the Earth's magnetic field and the atmosphere itself. The Earth's magnetic field curves between the north and south poles, with the most significant height between the magnetic field and the Earth's surface over the equator. Accordingly, the contribution of cosmic background radiation will increase with altitude and latitude. For this reason, aircrews are exposed to a greater level of background radiation, and astronauts are exposed to a still greater level.

Cosmic rays are responsible for the Aurora Borealis and the Aurora Australis, boreal being the Latin for northern, and austral being the Latin for southern.

Cosmic radiation interactions with the atmosphere produce a multitude of *cosmogenic* species. The most relevant are tritium and carbon-14. Both radionuclides are in an equilibrium state within our atmosphere; they are radioactively decaying at the same rate they are produced. The total global inventory for tritium is around $1.3 \times 10^{18}$ Bq, and carbon-14 is around $1.2 \times 10^{19}$ Bq.

### 7.2.1    Carbon Dating

Carbon dating relies upon the constant production of carbon-14 in the atmosphere. When an organism dies, it will stop exchanging carbon with the environment as it will no longer eat, breathe, or drink. At this stage, the carbon-14 within the organism will no longer be held in equilibrium with the environment and will be radioactively decaying at a predictable rate. Carbon-14 has a half-life of 5730 years, so the older the item you are attempting to carbon date, the fewer carbon-14 atoms will be present, and the accuracy of the calculated age will be diminished. This is why carbon dating cannot be applied to dinosaur fossils; they are too old.

## 7.3    Anthropogenic Background Radiation

The term anthropogenic refers to something that originates from human activity. The most notable sources of anthropogenic background radiation are fallout from atomic weapons testing, occupational radiation exposure, discharges from the nuclear and fossil fuel industry, and medical exposures.

The only significant anthropogenic background radiation is from medical exposures. The remaining sources of anthropogenic background radiation contribute less than 1% of the total background radiation. The medical exposures will include medical treatments, such as radiotherapy used for the treatment of cancers, and diagnostic exposures, which include everything from a dental x-ray to an abdominal computed tomography (CT) scan. A CT scan uses multiple X-rays to image 'slices' of a patient's body. As a diagnostic tool, they can be invaluable; however, multiple exposures will result in a more significant dose to the patient. Approximate doses for common diagnostic medical treatments are given in the table below:

| Medical procedure | Effective dose (mSv) |
| --- | --- |
| Dental X-ray (bitewing) | 0.005 mSv |
| Dental X-ray (panoramic) | 0.01 mSv |
| Chest X-ray | 0.1 mSv |
| Mammogram | 0.4 mSv |
| Head CT (Computed Tomography) | 2 mSv |
| Pelvic CT | 6 mSv |
| Abdominal CT | 8 mSv |

Radiotherapy uses radiation to kill cancerous cells preferentially. Resultantly, the doses received by patients must be high. A typical treatment plan for breast cancer will involve several exposures given throughout one to two months with an accumulated dose of 45 to 60 Gy. The justification for providing doses this high, of course, rests on the potential consequences of untreated cancers (Fig. 7.1).

### 7.3.1    Variability of Background Radiation

The average level of background radiation in the UK is 2.7 mSv. However, this figure is very variable. One of the main sources of background radiation is radon gas, which is found in higher concentrations in granite. Consequently, in areas that have a lot of granite bedrock, there will be higher levels of radon gas, and because of this, the average background radiation for most parts of Cornwall is around 7 mSv.

However, this variability is not just geographical; it is also temporal. In colder months, the levels of radon gas in people's homes increase. As radon seeps out of the ground,

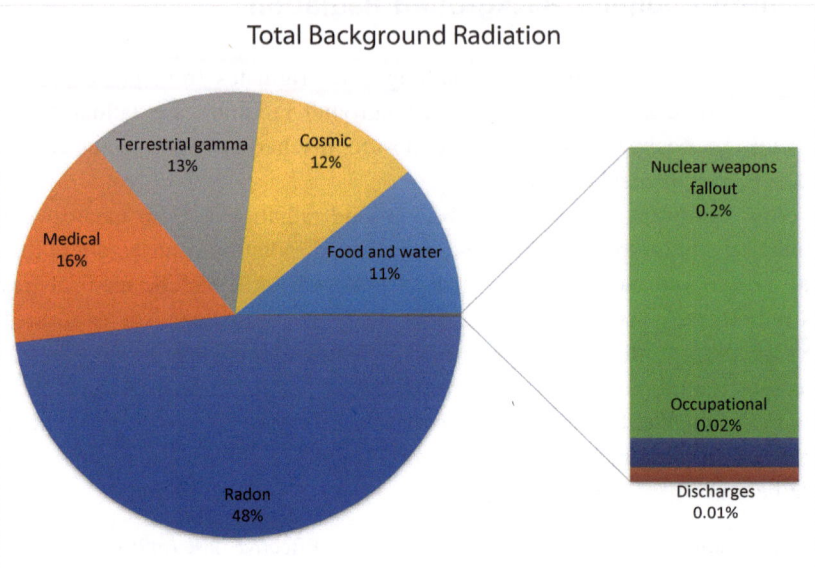

**Fig. 7.1** Total background radiation by source [8]

it can become trapped in houses. In colder temperatures, people keep their doors and windows closed, trapping more heat, and also more radon.

The global variability of background radiation is still more significant. Inhabitants in some areas of Ramsar, Iran, are exposed to 260 mSv/y of background radiation due to high concentrations of radium, thorium, and uranium in the rocks [9], and in coastal areas of Kerela, India, thorium-containing monazite sand results in dose rates of 70 mSv/y [10]. Whilst numerous studies have been done on these areas, a statistically significant correlation between higher cancer mortality and elevated background radiation has never been found. This is one of the most substantial pieces of evidence against the linear non-threshold model.

## 7.4　　Summary

- The three main categories of background radiation are primordial, cosmic, and anthropogenic.
- The three most important primordial isotopes are $^{235}$U, $^{238}$U, and $^{232}$Th, which create the actinium, uranium, and thorium decay series.
- Cosmic radiation originates from space and comprises high-energy particles. These particles can interact with molecules and atoms in our atmosphere to create other radioactive species.

- The largest source of anthropogenic radiation is from medical exposures.
- About 50% of background radiation in the UK is from radon gas
- The average background radiation in the UK is 2.7 mSv/y; this can vary and mainly depends upon the composition of the bedrock in the area.

## 7.5    Exercises

1. What are the three categories of background radiation?
2. Why is there so much focus on radon?
3. What was the isotopic concentration of $^{235}$U at the time of the Oklo reactors operation?
4. Why do aircrews experience increased levels of background radiation?
5. What percentage of background radiation in the UK is from medical exposures?

# Radioactive Waste Management

**8**

Radioactivity is everywhere: in the food we eat, the houses in which we sleep, and the air we breathe. Consequently, the waste that we produce will also be very minorly radioactive. Thus, the definition of radioactive waste cannot simply be 'waste that is radioactive'. To summarise the UK's *Environmental Permitting (England and Wales) Regulations of 2016*, radioactive waste is defined as no longer useful material containing a specific activity greater than certain thresholds [11]. It could come from creating new radioactive material, for instance, in the core of a nuclear reactor or from concentrating naturally occurring radioactive material.

## 8.1    Categories of Radioactive Waste

High-Level Waste (HLW)—is heat generating and therefore requires cooling. Examples include used nuclear reactor fuel rods.

Intermediate-level waste (ILW) is waste that contains more than 12 GBq/tonne ($\beta/\gamma$) or >4 GBq/tonne ($\alpha$). Examples include components from dismantled nuclear reactors located close to the core and either heavily contaminated or activated by the neutron flux.

Low-level Waste (LLW) is waste that contains <12 GBq/tonne ($\beta/\gamma$) or >4 GBq/tonne ($\alpha$). Examples include personal protective equipment, such as plastic gloves, overshoes, and protective suits, used in an area where radioactive contamination is present.

Very Short-Lived Waste (VSLW) is waste that contains short-lived radioisotopes. It mainly originates from research or medical facilities.

© The Author(s), under exclusive license to Springer Nature Switzerland AG 2025    47
J. Wain, *Ionising Radiation Protection*, Synthesis Lectures on Engineering, Science, and Technology, https://doi.org/10.1007/978-3-031-65525-8_8

## 8.2    Disposal Principles

There are three main principles for the disposal of radioactive waste.

*Dilute and disperse*—if a relatively small activity of material is diluted into a large part of the environment, for instance, the sea or the atmosphere, the overall effect will be intangible. The difficulty is ensuring the material remains diluted and that the overall quantity never exceeds a harmful level. In nature, various mechanisms can result in the accumulation of certain radionuclides, from specific estuarine sediments that bind with caesium compounds to filter-feeding shellfish that concentrate particulate radionuclides.

*Delay and decay*—the key advantage of radioactive waste, compared to other hazardous substances like arsenic or asbestos, is that the hazard will decrease as time passes. With short-lived radioisotopes, this can happen very quickly. The half-life of iodine-131 is eight days; thus, if we found ourselves suddenly in possession of a quantity of iodine-131, we could seal it very thoroughly and wait for it to radioactively decay and no longer be hazardous. After 80 days, ten half-lives would have elapsed, and only 0.098% of the initial activity would remain.

*Concentrate and contain*—Longer–lived radioisotopes must be safely stored for long periods. This is easier if you volumetrically reduce them. The UK previously dissolved used fuel elements in nitric acid, removed the valuable components, and concentrated the liquor in massive evaporators. After concentration, it was mixed with glass and allowed to cool in large stainless-steel containers that were robotically welded closed. These containers were placed in several further layers of containment.

## 8.3    Amount of Radioactive Waste

The Nuclear Decommissioning Authority (NDA) has forecast the total amount of radioactive waste produced by the UK from the start of the nuclear industry to 2140 as 4.7 million tonnes. 94% is LLW, 6% is ILW, and less than 0.1% is HLW. The total volume of HLW is 1000 $m^3$, roughly equal to a tennis court stacked 3.8 m high. For context, in 2022, the UK produced more than 300 million tonnes of waste, and 6 million tonnes of this was classed as hazardous waste.

Although the volume of HLW is low, it contains 95% of all radioactivity. Conversely, LLW contains only 0.01%.

## 8.4    Waste Activity Over Time

Radioactive waste contains many different isotopes, each with an associated half-life. The most hazardous radioactive waste is HLW, which is used fuel that has undergone many nuclear fissions. This used fuel will contain fission products, activation products, and neutron capture products. Some of these will have very long half-lives, for instance, $^{239}$Pu, with a half-life of 24,100 years. Consequently, some radioactive waste will remain hazardous for many thousands of years.

## 8.5    Disposal Options for Radioactive Waste

Disposal options for radioactive waste depend upon its concentration of radiation, specific chemistry, economic and political factors. Given below are some options for each waste category that are currently in use.

LLW has a high volume and a low concentration of radiation. LLW is transported to a LLW repository where it is supercompacted and placed into large metal containers similar to shipping containers. These containers are then filled with cement and placed into specifically designed concrete lined 'vaults'.

ILW is placed into containers, 500 L stainless steel drums or 3 m$^3$ stainless steel boxes, which are then filled with specific cement to immobilise their contents. Finally, the containers are placed in purpose-built storage facilities.

HLW may be stored in stainless steel containers in a specially designed, pH-controlled fuel storage pond or reprocessed. Reprocessing removes the uranium and plutonium, concentrates the remaining material, and turns it into glass, a process known as vitrification [12].

## 8.6    Summary

- The main categories of radioactive waste are high-level waste, intermediate-level waste, and low-level waste.
- Disposal principles include dilute and disperse, delay and decay, and concentrate and contain.
- Some types of radioactive waste will continue to present a hazard for thousands of years.

## 8.7    Exercises

1. Granite rock often contains uranium ore, so is it a radioactive waste?
2. What is the key difference between HLW and ILW?
3. What is the main advantage of the delay and decay principle of radioactive waste management?
4. Why will used fuel remain hazardous for a long time?
5. How might you dispose of $^{131}$I?

# Radiation Detection and Instrumentation

9

Our bodies cannot directly sense ionising radiation; thus, we need tools to sense it for us. Fundamentally, radiation instruments take ionising radiation and change it into something we can sense, from the click of a Geiger-Muller counter to the electronic display on a scintillator. There are many different mechanisms for doing this. Most commonly, an instrument will collect the free electrons created as part of the ionisation process and amplify them, creating either a continuous current or a pulse.

## 9.1    Ion Chambers

Ion chambers are the most basic design of instrument. The simplest ion chambers consist of two plates held at a potential difference; the electronics circuit will link the two plates with an ammeter and a power source to create the potential difference. Ion electron pairs are created between these plates by incident radiation, and the potential difference will drag the negatively charged electrons towards the positively charged anode and the ions toward the cathode. When the electrons reach the anode, they flow around the circuit, and this current is measured by the ammeter. The energy required to ionise an atom or molecule will depend upon how strongly bonded its electrons are to its nucleus. The average amount of energy absorbed in air per ionisation is 34 eV.

Consequently, in an ionisation chamber that uses air between the anode and cathode, we can calculate the number of electrons produced from varying amounts of energy deposition. If a photon was to deposit 400 keV in the space between these two plates, termed the detector's active area, then we would expect about 11,500 ($400 \times 10^3/34$) electrons to be created. If there were no potential difference across these plates the electrons and ions

J. Wain, *Ionising Radiation Protection*, Synthesis Lectures on Engineering, Science, and Technology, https://doi.org/10.1007/978-3-031-65525-8_9

would recombine; however, by applying a voltage we can draw these electrons and ions to their respective electrodes. Increasing the voltage will increase the force that acts upon them.

## 9.2    Neutron Detectors

Neutrons are uncharged, which makes them more complicated to detect. To overcome this, neutron detectors will typically convert the neutrons to something easier to detect. An example is the boron trifluoride detector. Within the detector, boron captures a neutron and becomes an unstable compound nucleus. This nucleus quickly disintegrates resulting in the production of an $\alpha$ particle:

$$^{10}B + n = {}^{7}Li + \alpha + \gamma$$

The creation of $\alpha$ particles within your detector overcomes the main challenge with $\alpha$ particle detection, avoiding the shielding effect of your detector housing. The energy these $\alpha$ particles deposit is converted to an electronic signal that is recorded by the instrument.

The efficiency of the detector will depend upon the fraction of neutrons that undergo capture reactions with the $^{10}B$ and this fraction will be dependent upon the neutron energy. Lower energy neutrons will have a greater probability of interaction; thus, neutron detecting instruments commonly include a moderator, often a plastic sphere. However, a neutron detector will not exhibit linear energy sensitivity even with a moderator. Low-energy neutrons may be absorbed by the moderator, and high-energy neutrons may not interact with the moderator or the $^{10}B$.

### 9.2.1    Geiger-Muller Detectors

The most common GM detector design consists of a hollow steel tube with a stainless steel or tungsten wire running down the middle. The tube and the wire are electronically insulated from one another, enabling the wire to act as the anode and the tube the cathode. There is a significant potential difference between the anode and cathode, typically around 400—900 V. There is often a thin mica window at one end of the tube to enable lower energy radiation to enter the tube. Within the tube, there is an inert gas at a low pressure, usually a mixture of helium, neon, or argon, and a quench gas, typically a halogen gas like bromine or chlorine.

When ionising radiation deposits energy within the tube or on the inside surface of the tube walls, it creates ionisations. The electrons produced from these ionisations will be pulled toward the anode. The closer they get to the anode, the more forcefully they are pulled towards it. As these electrons are accelerated, they will interact with more gas molecules, creating further ionisations. The electrons produced from these interactions will also be accelerated, resulting in a cascade of electrons. This cascade is known as a Townsend avalanche.

Additionally, these electrons will leave some of the gas molecules in an excited state. As they de-excite, they release a UV photon that creates further ionisations elsewhere in the tube. These photons travel at the speed of light, so the secondary ionisations they produce will occur practically simultaneously with the initial ionisations.

The electrons travel much faster than the ions, and the Townsend avalanches result in many ionisation events near the anode. Resultantly, there will be many ions near the anode. These ions will reduce the electric field generated by the significant potential difference and prevent further gas multiplication.

As all of the electrons hit the anode, they will produce a pulse of current within the instrument's electronics. The size of this pulse will be dependent upon the number of electrons that hit the anode, and because there are so many extra electrons generated by the UV photons and the Townsend avalanches, this pulse will be the same size every time; consequently, the instrument cannot discriminate between different energies of radiation, nor even different types of radiation. It is for this reason that GM detectors act as counters.

The quench gas is important to prevent the instrument from remaining saturated with ions and electrons and to bring the detector back to its initial start state. The time required for this is dependent upon the time required for the ions to slowly migrate towards the cathode and regain an electron. This insensitive time is termed *dead time*.

A higher count of pulses indicates a greater number of ionisations within the tube, which is displayed as counts per second (cps).

## 9.3   Scintillators

A scintillator is a material that scintillates, releases light, after energy is deposited within it. When ionising radiation deposits energy within a scintillator, it creates excited states within that material. These excited states de-excite by emitting photons. The scintillators used in instruments do this quickly in a process termed fluorescence. These photons are collected in a photomultiplier tube (PMT). Within the PMT, the photons are converted to electrons in a photocathode; the electrons are then focused onto dynodes where electric potential differences accelerate and multiply them. At the end of the dynode chain, the electrons are collected by the anode and a pulse of current travels from the scintillator probe into the rest of the instrument.

The size of this pulse will be proportional to the number of electrons collected by the anode, which is proportional to the number of photons that were converted to electrons, which is proportional to the amount of energy deposited in the scintillator. Thus, scintillators can differentiate between different types of radiation that deposit different amounts of energy. A gamma photon will typically have significantly less energy than an alpha particle. Additionally, a gamma photon is unlikely to deposit all its energy in one location. If both were to enter the active volume of a scintillator, the pulse produced by the alpha particle would be much larger. The electronics within the instrument may then count the number of big pulses and the number of smaller pulses, indicating the number of alpha particles detected and the number of gamma photons detected.

An ideal scintillator material will be transparent to the wavelength of light it creates, have a high efficiency of converting incident radiation into photons, create photons that can be easily converted into electrons by the photocathode, and rapidly create the photons. This perfect material is yet to be found and some materials work better with certain types of radiation. Some examples of inorganic scintillator materials include ZnS for alpha detection, NaI, and CsI. Some examples of organic scintillator materials include anthracene and stilbene crystals.

Some scintillator materials can be dissolved in a solution for use as a liquid scintillator. A radioactive sample is mixed with this solution, surrounding a radioisotope with scintillant. This method enables the detection of low-energy beta emitting radioisotopes, such as tritium and $^{14}$C.

## 9.4    Semiconductors

### 9.4.1    Semiconductor Theory

Individual atoms have orbital electrons that are bound in discrete energy levels. From the nucleus, these energy levels are labelled k, l, m, n, o... shells. The most stable position for the electron is the shell closest to the nucleus, the k shell. At this position, the force between the negatively charged electrons and the positively charged protons is the greatest. To move to a higher orbital the electron will require additional energy.

When atoms are spaced close together as they are in a solid, the atoms will affect those around them. The individual energy levels will coalesce into energy bands that extend throughout the solid. The two most important bands are called the valence band and the conduction band. The valence band relates to the electrons that occupy the outermost energy level in an atom, the valence shell. The conduction band is the next energy band above this, in terms of energy. Electrons in the conduction band can flow through the solid and the flow of electrons is an electric current, thus the position of the conduction band relates to the ability of the solid to conduct electricity. In different solids, these bands will have different energies, and importantly, the gap between these two bands will be different. Electrons can move through this band gap but can never come to rest within it. Materials without a band gap are termed *conductors* as it is easy for the electrons to move from the valance band into the conduction band and thus the material is electrically conductive. Copper is a good example of a conductor. Materials with a large band gap are termed insulators and materials with a medium band gap are termed semi-conductors.

The application of a potential difference to a semi-conductor will drag the electrons towards the anode, and drag holes, locations that are no longer occupied by electrons, toward the cathode. These holes represent the absence of a negative charge and behave as though they were a point positive charge. For an electron or a hole to move throughout a semiconductor a single electron will move to an adjacent position and the electron that was occupying that position will do the same, and so on. This can be imagined like a Newton's cradle, one ball swinging into a line of balls that transmit the effect to the very end ball. Consequently, semiconductors are some of the fastest-responding detectors.

When ionising radiation deposits energy within a semiconductor many ion–electron pairs will be produced along its track. The main advantage of semiconductors is the low amount of energy required to produce each pair. For Si or Ge, this value is around 3 eV, about ten times less than in air. Thus, the number of charge carriers is ten times greater. This dramatically increases the achievable energy resolution.

Germanium and silicon are the two most common semiconductors. They both have four electrons in their valence electron orbital. Atoms with eight valence electrons have a full outer shell of electrons and are more chemically stable. Consequently, germanium and silicon will be more stable if they gain an additional four electrons. Using germanium as an example, within a pure crystal it will achieve a full outer shell by sharing four electrons with the other germanium atoms around it.

Germanium has a low bandgap of about 0.7 eV. Consequently, these detectors must be cooled to reduce the number of thermally generated electrons moving from the valance band into the conduction band and diminishing the energy resolution of the detector. This is achieved by mounting the HPGe detector in a vacuum chamber that is attached to a liquid nitrogen dewar.

## 9.5 Personal Dosimeters

### 9.5.1 Thermoluminescent Dosimeters (TLDs)

The term thermoluminescent describes a material that is capable of emitting light in response to an increase in temperature. This section will focus specifically on TLDs as used commonly throughout the nuclear industry for personal dosimetry. These TLDs are typically made from LiF crystals which have been 'doped' with impurities, including Mg and Ti. Ionising radiation will create free electrons. The liberated electrons will move from the valence band, into the higher energy conduction band. In pure LiF these electrons would simply drop back down into the valence band, recombining with the holes created by other liberated electrons. However, by doping the LiF crystal we can create electron traps, metastable states where the electrons do not have quite enough energy to drop to the valence band.

After specific intervals, usually every one or three months, the dosimeters are analysed. They are slowly heated in a process known as annealing. As a result of the temperature increase, the individual atoms within the crystalline structure gain more energy. As the energy continues to increase the electrons that were displaced from the valence band can overcome the small energy barrier created by the metastable state and fall back to the valence band, losing energy as they do so. The excess energy that they lose is released as a photon. The energy, and thus wavelength, of this photon is dependent upon the band gap within the crystal. The number of photons released correlates to the number of electrons released from the metastable states, which correlates to the initial amount of radiation deposited in the TLD.

As the temperature is increased further the crystalline structure will return to its lowest energy state and thus the TLD can be reused many times.

LiF TLDs exhibit a minimum sensitivity of about 100 μGy and have a fairly linear response up to 4 Gy.

## 9.6    Cherenkov Radiation

Cherenkov radiation occurs when a charged particle travels faster than the speed of light in a dielectric medium. It is named after the Soviet scientist Pavel Alekseyevich Cherenkov, who was awarded the Nobel Prize in Physics in 1958 for its discovery. Cherenkov radiation is the ethereal blue glow often seen in photos of nuclear reactors that are submerged in water.

Charged particles can travel faster than the speed of light in a medium because some mediums will slow down light more than they will slow down a charged particle. The most common example is high-energy electrons travelling through water.

A dielectric material does not conduct electric current easily and can be polarised by an electric field. Water is a polar molecule, meaning it has a permanent electric dipole moment due to the unequal sharing of electrons between the oxygen and hydrogen atoms.

As the charged particle moves through the dielectric medium it will polarise the atoms or molecules within the medium, asymmetrically. This polarisation will slightly displace the electrons (within the medium) from their normal positions. As these electrons relax to their lower energy, initial, positions they release their excess energy as a photon. As the charged particle is moving faster than the speed of light in the medium, this creates constructive wave interference, from the asymmetrical polarisation. It is this constructive interference that produces the characteristic Cherenkov light. Cherenkov radiation can be imagined as a sonic boom from a jet flying faster than the speed of sound.

The relativistic kinetic energy for a high-speed electron may be calculated with:

$$KE_R = m_0 c^2 \left[ \frac{1}{\sqrt{1 - v^2/c^2}} - 1 \right] \qquad (9.1)$$

where:

$KE_R$   is the relativistic kinetic energy of a high-speed electron.
$m_0$   is the rest mass of the electron.
c   is the speed of light in a vacuum.
v   is the velocity of the electron.

The refractive index in a medium may be calculated with:

$$n = c/p_v \qquad (9.2)$$

Where:

$p_v$   is the speed of light in that medium (the phase velocity).

The threshold energy for Cherenkov energy to occur will be when the KE of the electron is equal to the speed of light in that medium, when $KE_R = p_v$. Substituting $c / n$ for v and 0.511 MeV for $m_0 c^2$ in Eq. 9, the threshold energy (in MeV) for a high-speed electron is:

$$KE_R = 0.511 \left[ \frac{1}{\sqrt{1 - 1/n^2}} - 1 \right]$$

The refractive index for water for visible light is about 1.33. Consequently, we find the threshold energy for a high energy electron in water for visible light is 0.264 MeV.

Cherenkov detectors utilise a detecting medium and a photomultiplying tube to amplify, and enable the electronic processing of, the signal. They are typically confined to high energy physics experiments; however, they can be used in liquid scintillation, as long as the source to be measured emits beta particles with an energy exceeding the Cherenkov threshold.

## 9.7     Calibration

Radiation instruments must be calibrated frequently if they are used operationally. Calibration is performed by exposing an instrument to a known level of radiation and ensuring it responds as expected. It is performed in specialist calibration facilities using carefully controlled radiation fields traced to national standards.

## 9.8     Minimum Detectable Amount (MDA)

Confirming or denying the presence of low levels of radiation can be difficult. Ubiquitous background radiation will ensure that whenever you measure a sample, your radiation instrument detects background radiation in addition to radiation that may or may not be on your sample. Consequently, when measuring a sample, you must first measure the background radiation, B, and then subtract this value from your measurement of the sample, S + B:

$$(S + B) - B = S \qquad\qquad (9.3)$$

If there was no variability in background radiation, and your detector was perfect, you could confidently assert that if S > 0, the sample contains radioactivity. However, there will be fluctuations in the background radiation detected and, unfortunately, your detector will not be perfect. The result will be many false positives. Thus, we must determine a value to record as a positive result, P. This quickly becomes a balancing act. The higher you set this value the fewer false positives you have, but equally, the higher you set this value the more low-level samples you will miss, false negatives.

Radioactive decay is an intrinsically stochastic process. If you designed an experiment where your instrument is detecting around 50 counts of background radiation, and you repeated the experiment infinite times, the number of counts you would detect after each experiment would follow a Gaussian distribution, centred on the mean of 50 counts. Most of the time you would detect 50 counts, but sometimes you would detect more, and sometimes fewer (Fig. 9.1).

A Gaussian distribution is a type of probability distribution that is symmetric and bell-shaped. It is characterized by a specific mathematical formula and is commonly used in statistics to model the distribution of random variables. The probability density function of a Gaussian distribution is given by:

$$f = \frac{1}{\sqrt{2\pi\sigma^2}} e^{-\frac{(x-\mu)^2}{2\sigma^2}}$$

Where:

**Fig. 9.1** An example Gaussian distribution from multiple detector measurements

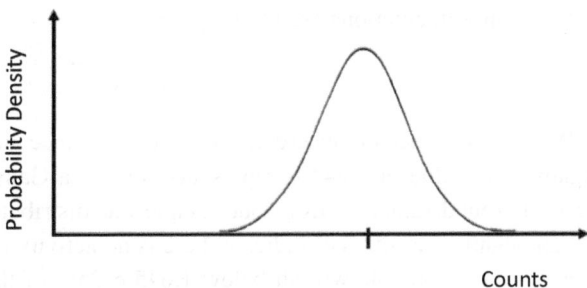

f    is the probability density function.

x    is the variable.

μ    is the mean value of the distribution.

$\sigma^2$    is the variance (the measure of spread).

σ    is the standard deviation.

Within a Gaussian distribution, one standard deviation will incorporate 68% of all values, two standard deviations 95%, and three 99.7% (Fig. 9.2).

To determine the standard deviation for the sample, $\sigma_S$, we must combine the standard deviations associated with the sample and the background radiation, $\sigma_{S+B}$, and the standard deviation associated with the background radiation, $\sigma_B$:

$$\sigma_S = \sqrt{\sigma_{S+B}^2 + \sigma_B^2}$$

If there is no activity on the sample then this becomes:

$$\sigma_S = \sqrt{2\sigma_B^2}$$

$$\sigma_S = 1.414\sigma_B$$

**Fig. 9.2** Gaussian distribution

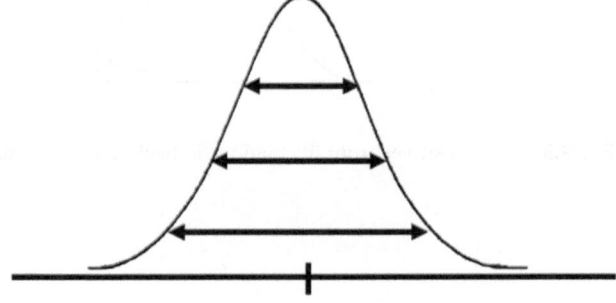

If the only fluctuations are random then:

$$\sigma_B = \sqrt{C_B}$$

We can now decide where to strike our balance between false positives and false negatives. A value of 1.645 σ represents 90% of a Gaussian distribution. By subtracting the background radiation from our sample our distribution will be centred around zero, and all negative results will indicate there is no activity on our sample. Thus, 1.645 σ will mean a random sample will sit below 1.645 σ 95% of the time. Consequently:

$$P > 1.645 \times 1.414\sigma_B$$

The result will be a 5% chance of attaining a false positive.

However, if the sample contains radioactivity then this will also be expressed as a Gaussian distribution. Consequently, we must further offset P. The correct value for P will now be dependant upon specific operational factors that reflect the tolerable number of false positives or false negatives.

In the diagram below P shows the number of counts chosen by the operator to be treated as a positive sample, the two Gaussian distributions from background radiation and the sample are show. The mean value of counts on the sample will change as samples are more or less radioactive. The shaded area in red shows the region where false positives may occur (Fig. 9.3).

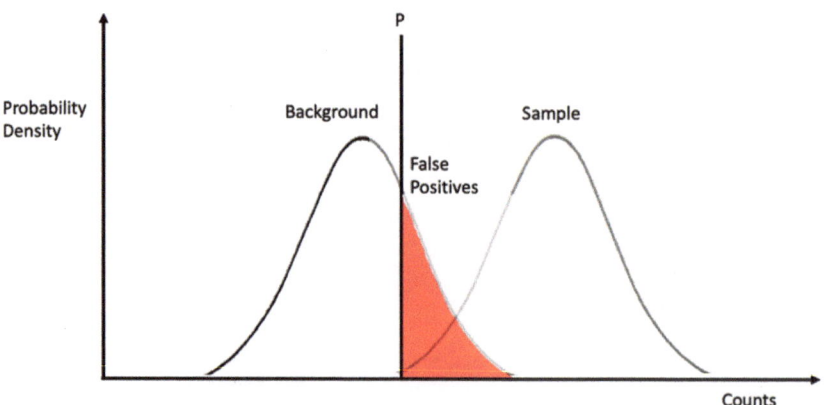

**Fig. 9.3** False positives from fluctuations in background radiation

### 9.8.1  Solid Angle, $\Omega$

The solid angle is a method of describing an angle in 3D as a cone projected away from the radius of a sphere.

In 2D, we may use degrees, with $360°$ in a circle, or we can use the units of radians, with $2\pi$ radians in a circle. A radian is a pure mathematical concept. By taking the length of the radius of a circle and measuring the same length around the edge of the circle, you can create a wedge. The angle of this wedge will be the same for any sized circle, and there will be $2 \times \pi$ wedges in a circle. We can extend this idea into 3D using the units of the steradian, with $4\pi$ steradians in a sphere. One steradian is a cone projected from the centre of a sphere onto its surface. The 3D angle of this cone is always such that the area of the base of the cone projected on the surface of the sphere is $r^2$. As there are $4\pi$ steradians in a sphere, the surface area of a sphere is $= 4\pi r^2$.

Within the field of nuclear instrumentation, the steradian enables us to calculate the amount of radiation incident upon an instrument at different distances and geometric configurations.

## 9.9   Summary

- Ion chambers apply a potential difference across two surfaces to draw electrons and ions to an anode and cathode, respectively. When electrons are collected by the anode, they increase the current flowing around the circuit, which is measured by an ammeter.
- Neutrons are uncharged, so we use intermediatory reactions to absorb them and emit something easier to measure.
- Geiger-Muller counters apply a significant potential difference between an anode and a cathode. This potential difference is large enough to accelerate the electrons produced from ionising radiation to cause further ionisations and to excite atoms / molecules within the detector. The excited atoms / molecules deexcite by the emission of a UV photon, which causes additional ionisations.
- Scintillation detectors produce photons in response to energy deposition within their active volume. By adding 'doping agents' to the scintillant, the wavelength of this light can be altered. This light is collected by a photomultiplier tube, and an electrical signal is created.
- Semiconductors are materials with an intermediate band gap between their valence and conduction bands. Electrons move through semiconductors quickly by displacing one another. The low ionisation energy enables a higher energy resolution and spectroscopic information to be collected.
- Thermoluminescence Detectors (TLDs) comprise of a doped crystalline structure. After irradiation, they can be slowly heated up and will produce light at an intensity that correlates to how much energy has previously been deposited within them.

- Cherenkov radiation is responsible for the blue glow observed around high-activity sources in water. Higher-density transparent mediums slow light, but the speed of electrons are less affected. When electrons start moving faster than the speed of light in a substance, Cherenkov radiation is produced.

## 9.10    Exercises

1. What energy band gap will create visible light?
2. What are the main advantages of semiconductor detectors?
3. What is the main advantage of a GM tube?
4. List five advantages of TLDs.
5. How many steradians makeup half a sphere?

# Shielding

# 10

There are many different sources of radiation, from fission products to radon gas. Different types of radiation correlate with different types of hazards, and it is common to find radiation sources that produce multiple types of radiation. In this chapter, each type of radiation is dealt with in turn to enable the reader to understand the most important considerations for any scenario.

## 10.1    Alpha

Alpha particles are identical to helium ions, consisting of two neutrons and two protons, and thus have a charge of 2+. Once an alpha particle gains two electrons, it will become an atom of helium gas. The large difference in mass between an alpha particle and an electron results in poor energy transfer; thus, alpha particles travel in straight lines through matter. Rarely, alpha particles interact with an atomic nucleus and are deflected by a large angle, the target nucleus gaining significant energy. Accordingly, the energy loss rate within a shielding material will depend upon the ratio of electrons to nucleons, and counterintuitively, alpha particles will travel further in high atomic number materials, like lead, than low atomic number materials with the same density.

Alpha particles typically have high energy, around 5 MeV, and are produced with discrete energies, making alpha spectrometry possible. The alpha particle's large charge results in intense ionisation, causing rapid energy loss. A typical range in air for a 5 MeV alpha particle is around 4 cm. As an alpha particle begins to slow down, the effect of its large charge will increase, slowing it down faster. This results in an increase in stopping

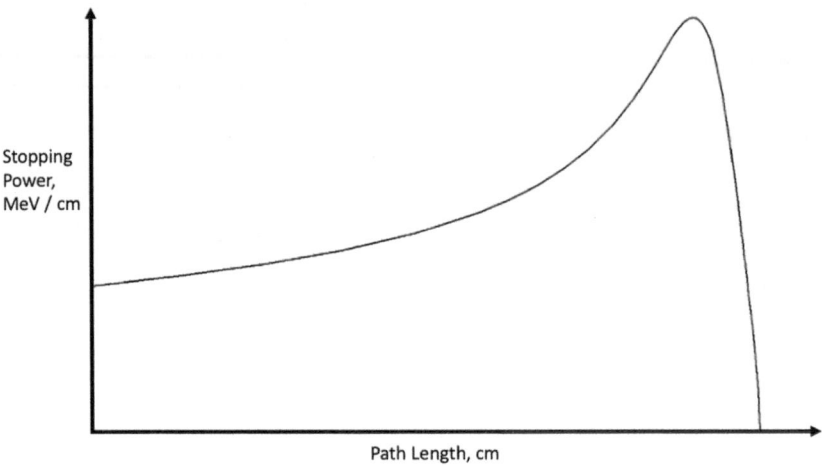

**Fig. 10.1** Bragg curve of a typical alpha particle

power, measured in MeV / cm. The stopping power of a typical 5 MeV alpha particle will be 2.5 MeV / cm. This can be seen in the Bragg curve in Fig. 10.1.

When this high energy is combined with their short range, a large amount of energy will be deposited per unit mass. Biologically, the entirety of an alpha particle's energy may be deposited within a few cells. This creates a sizeable internal hazard.

## 10.2   Beta

Beta particles are electrons and are created simultaneously with an antineutrino. The energy is divided between the beta particle and the antineutrino, resulting in a spectrum of beta particle energies with a defined maximum possible energy. Beta particle creation can be thought of as a neutron turning into a proton and a high-energy electron. The proton number determines the element, so as the proton number changes, the element will change:

$$^{14}C \rightarrow \ ^{14}N + \beta + antineutrino$$

Beta particle energies are quoted as either average energies or maximum energies. Radiation protection focuses on maximum energies. Typically, the range in air of beta radiation is around 3 m; however, this is not well defined. Orbital electrons easily deflect beta particles; thus, the path of a beta particle will zigzag erratically. The maximum energy of beta particles will vary by isotope. For instance, tritium has a maximum beta particle energy of 18.6 keV, whereas $^{90}Y$ decays with an energy of 2.3 MeV. Consequently, tritium requires

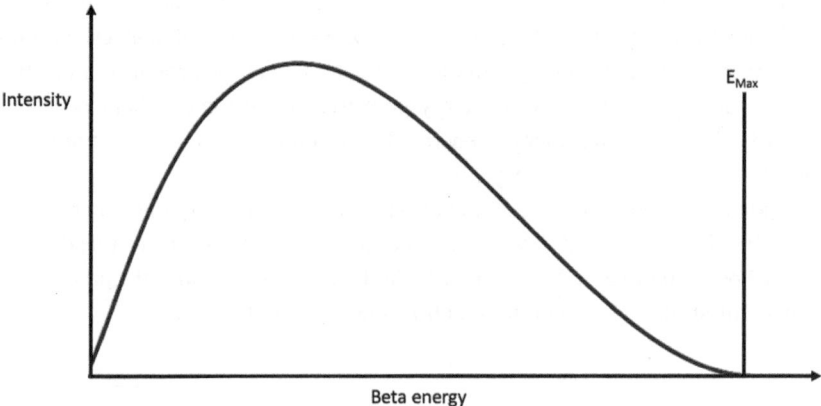

**Fig. 10.2**  Spectrum of beta energies at emission

specialist liquid scintillation instrumentation to be detected, and $^{90}$Y's betas have a range in air of around 9 m (Fig. 10.2).

### 10.2.1  Bremsstrahlung Radiation

Bremsstrahlung radiation, literally 'breaking-radiation', describes the photons released as charged particles rapidly accelerate or decelerate in the presence of other charged particles. In radiation protection, this is most significant for high-energy electrons decelerating in the presence of atomic nuclei. These high-energy electrons can be produced either through beta particle generation or by applying a high potential difference across an anode and cathode. When photons are created in this manner, they are defined as x-rays. The amount of energy released as Bremsstrahlung radiation depends upon the initial energy of the incident particle, the mass of the incident particle, and the charge of the surrounding static particles. An incident particle with a large mass will have more momentum. It thus will accelerate or decelerate slower than a lighter particle, and if the surrounding particles have a significant charge, they will exert more force upon the incident particle. The fraction of energy converted to x-rays by the deceleration of electrons may be approximated by:

$$E_x = {}^{EZ}\!/_{3000} \tag{10.1}$$

where:

E is the maximum energy of the incident beta particle in MeV.

Z is the atomic number.

Using the above expression, the maximum energy of beta particles produced from $^{32}$P is 1.71 MeV; when interacting with lead (Z = 82), it will convert around 4.7% of its energy into bremsstrahlung radiation. The total energy produced may be calculated by multiplying the fraction of energy converted by the mean beta particle energy (0.695 MeV). Thus, the total energy generated is around 32 keV. Bremsstrahlung energy conversion is clearly a low-efficiency process.

The avoidance of bremsstrahlung radiation generation is frequently achieved with a composite shield. A lining of low atomic number material, such as plastic, is placed within a lead pot. The plastic lining shields the beta particles and only produces a small amount of bremsstrahlung radiation, which is shielded by the lead.

## 10.3  Photons

### 10.3.1  Half Value Layer

HVLs are the thickness of the shield that will reduce the dose to half of the initial value; they are closely related to Tenth-Value Layers (TVLs), that is the thickness required to reduce the dose to 1/10 of the initial value. The HVLs will depend upon both the shielding material and the energy of the incident radiation. The use of HVLs include approximations but generally form reasonable estimations for shielding calculations. Plotting the decrease in radiation as it passes through a shield using HVLs will show an exponential decrease, analogous to a plot of radioactive decay. The following table contains typical values for HVLs at various energies of gamma radiation:

| Energy, MeV | Approximate HVLs, mm | | | |
|---|---|---|---|---|
| | Lead | Iron | Concrete | Water |
| 0.5 | 5 | 10 | 33 | 8 |
| 1 | 8 | 15 | 46 | 10 |
| 1.5 | 13 | 18 | 58 | 12 |
| 2 | 15 | 20 | 66 | 15 |

A valuable equation to help use the concept of HVLs is:

$$D_1 = D_0(0.5)^{x/HVL} \tag{10.2}$$

where:
  $D_1$—dose after the shield.
  $D_0$—dose before the shield.
  X—thickness of the shield.

## 10.3.2  The Interaction of Photons with Matter

The predominant photon interaction, from the photoelectric effect, Compton scattering and pair production, will depend upon the atomic number of the shielding medium and the energy of the incident photon. Each interaction has an associated cross-section per atom, $\sigma$. Photon shielding can be thought of in terms of the number of photons of a specific energy, the total number of photons, or the total energy of photons that remain after a shield.

The photoelectric effect is dominant at low photon energies below 0.5 MeV. Here, an incident photon transfers all its energy to an orbital electron. The amount of energy transferred exceeds the binding energy of the electron, and the electron is stripped from the atom, creating an ion. The photoelectric cross-section depends upon the number of target electrons and the energy of the incident photon. It increases rapidly as the photon energy decreases. It is most significant with high atomic number elements and low photon energies. The photoelectric cross-section is approximately proportional to:

$$\sigma_{PE} \propto E_{\gamma}^{-7/2} \qquad (10.3)$$

$$\sigma_{PE} \propto Z^{3-5}$$

where:

$E_{\gamma}$ is the energy of the incident photon.

$Z^{3-5}$ relates to the atomic number to the power of a value between three and five, dependent upon the photon energy and the element.

At intermediate photon energies, Compton scattering is dominant. Here, an incident photon transfers some of its energy to an orbital electron. The electron is ejected from the atom, and another photon with a lower energy is produced. The Compton scattering cross-section per atom is proportional to the atomic number of the material and is inversely proportional to the energy, in the range of 0.5–2 MeV:

$$\sigma_{CS} \propto Z$$

$$\sigma_{CS} \propto {}^{1}\!/_{E_{\gamma}}$$

The angle of the secondary photon is dependent upon its energy. Higher energy photons are forward scattered, so they maintain a trajectory similar to the initial incident photon. We may calculate the energy of the Compton scattered photon:

$$E_{\gamma'} = \frac{E_{\gamma}}{1 + \left(E_{\gamma}/m_e c^2\right)(1 - \cos\theta)}$$

where,

$E_{\gamma'}$ is the energy of the Compton scattered photon.

$E_\gamma$ is the energy of the initial photon.

$m_e$ is the rest mass of an electron.

c is the speed of light.

$\theta$ is the angle of scatter of the photon.

The scattering angle can range between $0°$ indicating no scatter and no transfer of energy, and $180°$ indicating backscatter and a maximum energy transfer.

Compton scatter and the photoelectric effect are not necessarily exclusive. For instance, a high-energy photon is likely to have multiple Compton scattering interactions before being photoelectrically absorbed.

At high photon energies, pair production is dominant. Pair production is an Einsteinian energy to mass transformation. When near an atomic nucleus, a photon with energy exceeding 1.022 MeV can transform into an electron and a positron. The strong electric field encountered by the photon in the vicinity of an atomic nucleus is required to maintain the conservation of momentum. A positron is antimatter, an anti-electron, and will annihilate when it interacts with another electron. The product of this annihilation will be two gamma photons, emanating at a perfect $180°$ from one another, each with an energy of 511 keV. These precise energies result from the energy to mass conversion of an electron according to $E = mc^2$. The pair production cross section is approximately proportional to:

$$\sigma_{PP} \propto Z^2$$

The total cross-section for photon interactions is:

$$\sigma_T = \sigma_{PE} + \sigma_{CS} + \sigma_{PP}$$

When considering the number of photons shielded by a medium, every photoelectric interaction and pair production will reduce the number of photons by one. However, some secondary, lower energy photons produced from Compton scattering will pass through the shield. Consequently, only a fraction of the Compton scattered photons will be ultimately removed.

The microscopic cross-section, $\sigma$, is used frequently in neutron shielding; however, it is often more common to use the closely related concept of linear absorption coefficient, $\mu$, with units of $cm^{-1}$, in the field of photon shielding:

$$\mu = N\sigma_T$$

where:

N is the number of atoms per cubic centimetre.

When comparing two shielding materials, it can be advantageous to consider how much mass each material requires. This can be achieved by the use of the mass absorption coefficient, $\chi$, in units of cm$^2$ g$^{-1}$:

$$\chi = \mu/\rho$$

where:
  $\rho$ is the density of the material in g cm$^{-3}$

The average distance travelled by a photon in a material is defined as the mean free path, mfp:

$$mfp = {}^{1}/_{\mu}$$

The rate of interaction in the target is thus given by:

$$R = \mu\phi_{\gamma}Ax$$

where:
  $\varphi_{\gamma}$ is the photon flux, in photons cm$^{-2}$ s$^{-1}$
  A is the area.
  x is the thickness of the shield.

Finally, the attenuation of a collimated beam of photons after a certain thickness of shield may be estimated by:

$$\phi_x = \phi_0 e^{-\mu x}$$

where:
  $\varphi_x$ is the photon flux after x thickness of shielding.
  $\varphi_0$ is the unshielded photon flux.

However, many sources of photon radiation are not collimated beams. By assuming they are isotropic point sources, infinitely small radiation sources emanating photons in a perfect sphere all around them, we can account for distance variations. As the radiation is produced spherically, an increased distance from the source will reduce the number of photons per area by an inverse square relationship. The surface area of a sphere of radius, r, is given by:

$$SA = 4\pi r^2$$

Consequently,

$$\phi_r = {}^{S}/_{4\pi r^2}$$

where:
  S is the total photon flux.
  $\varphi_r$ is the photon flux at a distance r.

Thus, we can account for geometric and shielding effects:

$$\phi_r = \frac{S}{4\pi r^2} \cdot e^{-\mu x}$$

This equation is a reasonable approximation when close to a physically small source, less than 10 m, and with thin shielding, only a few mean free paths. As we move further from the source the effect of air attenuation will increase and as the thickness of the shield increases the effect of build-up will become significant.

### 10.3.3  Build up

In thicker shields, the number of photons reaching the far side can be greater than those at the exact location if no shield is in place. This is because of Compton scattering interactions. In thicker shields, a single photon may undergo multiple Compton scatters. Consequently, the total photon flux will not decrease exponentially, as would be true if only the photoelectric effect and pair production were considered. It is for this reason that HVLs are only an approximation. When the incident radiation is not monoenergetic, lower energy photons will be attenuated more quickly than higher energy photons. Additionally, when higher energy photons undergo Compton scattering, they are forward scattered, so more will maintain their initial trajectory and reach the far side of the shield. The result will be a filtering effect of the lower radiation energies that causes the overall beam of radiation to be of an average higher energy after leaving the shield, termed *beam hardening*.

To account for build up, we can use build up factors, such as B. These are defined as the ratio of some measurable quantity of the beam (flux, energy, dose etc.), with total quantity divided by the uncollided quantity.

$$\phi_r = \frac{S}{4\pi r^2} \cdot e^{-\mu x} \cdot B$$

The energy spectrum of the photons after the shielding will be markedly different. Operationally, radiation detectors will be more sensitive to specific energies of radiation. Thus, a radiation detector may appear to accentuate the effects of build-up.

### 10.4    Neutrons

Neutron shielding presents the most complex shielding challenges. Neutrons may be either scattered or absorbed by nuclei. The scattering interactions involve the neutron transferring some or all of its energy to the nucleus but remaining free afterwards. Scattering may be either elastic or inelastic and will sequentially reduce the neutron's energy until it is absorbed or thermalised. After a neutron is finally captured by a nucleus, a gamma photon will be released, contributing to the overall shielding problem. Each will have a different interaction probability known as a cross-section. This cross-section will change with the neutron energy, and the neutron energy will be constantly changing after each interaction. The cross sections are not simple curves and will have erratic zones created from quantum resonances.

### 10.4.1  Sources of Neutrons

A nuclear reaction is summarised using the following notation: *target nucleus (ingoing particle, outgoing particle) product nucleus*. This notation will be used throughout this section.

There are two primary sources of neutrons: radioisotope sources and fission sources. A radioisotope source will consist of a light nucleus, typically Be or B and an $\alpha$ emitter. A common neutron-producing radioisotope source is $^{241}$Am$^9$Be, the Am being an alpha emitter with a 432-year half-life:

$$\alpha +^9 Be =^{12} C + n$$

Using the above notation, this is a $^9$Be$(\alpha, n)^{12}$C reaction.

A fission source describes neutrons produced as fission takes place. This may be spontaneous fission, as is the case with $^{252}$Cf, which has a half-life of 2.65 years and produces 4 neutrons, or it may be from neutron-induced fission, as with $^{235}$U.

### 10.4.2  Neutron Energy Groups

As their energy governs the behaviour of neutrons they have been split into loosely defined energy groups named:

| Name | Energy range | Description |
| --- | --- | --- |
| Thermal | 0.025 eV | Neutrons that are in thermal equilibrium with their surroundings, usually defined at 20°C |
| Epithermal | 0.2 eV | Neutrons with energies slightly exceeding those of thermal energies |
| Cadmium | <0.4 eV | Neutrons that are strongly absorbed by Cadmimum |
| Slow | 1–10 eV | |
| Resonance | 1–300 eV | Usually refers to neutrons that are strongly captured in the resonances of $^{238}$U |

(continued)

(continued)

| Name | Energy range | Description |
|------|--------------|-------------|
| Intermediate | 100 eV–0.5 MeV | |
| Fast | 0.5 MeV | |
| Fission | 100 keV–15 MeV | The spectrum of neutron energies created from fission reactions |
| | (2 MeV average) | |

### 10.4.3  Elastic Scattering (n, n)

This may be imagined as a classic billiard ball interaction. Kinetic energy is transferred from, or to, a neutron. If the energy is transferred from the neutron to the nucleus, the target nucleus remains in its ground state. This can create a highly energetic recoil nucleus that transfers energy to the surrounding material via ionisation and excitation. Elastic scattering is essential in the moderation of neutrons: the term moderation refers to the reduction of energy of the neutron. The neutrons will be moderated until they are in thermal equilibrium with their surroundings; they will then be termed thermal neutrons. The fractional transfer of energy will depend on the target nucleus's mass and the angle at which the neutron is scattered. The most significant transfer will occur when the mass of the target is as close to the mass of the neutron as possible and the neutron is backscattered. The minimum fractional energy after one scatter, $\propto$, is approximately equal to:

$$\propto = \left( {}^{A-1}/_{A+1} \right)^2$$

where:
A is the atomic mass of the target nucleus.

### 10.4.4  Inelastic Scattering (n, n′)

This is similar to elastic scattering; however, some of the neutron's energy causes excitation of the target nucleus. The nucleus will subsequently de-excite, emitting a characteristic gamma photon. Inelastic scattering occurs predominantly with fast neutrons.

### 10.4.5 Radiative Capture Reactions (n, λ)

This is where a nucleus absorbs a neutron and becomes a short-lived excited compound nucleus. This compound nucleus then de-excitates by the emission of a particle or photon, γ emission is just one of the de-excitation mechanisms. The photon will have a characteristic energy dependent upon the nucleus. Photon emission is common with slow neutrons, as particles require more energy to overcome the Coulomb barrier of the nucleus.

Radiative capture reactions are widely used in neutron detectors. An example is a $BF_3$ detector that uses a $^{10}B(n,\alpha)^7Li$ reaction.

### 10.4.6 Moderation

Common moderators include hydrogen, deuterium ($^2H$), and carbon. Their properties are shown in the graph below:

| Moderator | Mass | Average number of collisions to thermalise | Material |
|-----------|------|--------------------------------------------|----------|
| Hydrogen  | 1    | 18                                         | Water    |
| Deuterium | 2    | 25                                         | Heavy water |
| Carbon    | 12   | 114                                        | Graphite |

It is clear that hydrogen gives the best moderation; however, water will also absorb neutrons at a greater rate than deuterium or carbon. These are some of the considerations when designing moderators for nuclear reactors.

### 10.4.7 General Principles of Neutron Shielding

Most neutron shields are composite shields, initially a low atomic number material is used to moderate the neutrons, then a secondary material is used to absorb the neutrons, a final material is used to shield the γ photons produced by the various neutron interactions. These different materials can be layered or incorporated into one substance. An example of a composite shield of this kind may be polythene, boron, and lead.

## 10.5 Chapter Summary

- α particles have a high energy, high mass and high charge. They travel in short, straight lines unless they interact with an atomic nucleus. They present minimal shielding problems.

- β particles interact with other electrons, creating an erratic path that extends around 3 m from the source. When electrons are rapidly slowed, they can create secondary *bremsstrahlung* radiation.
- Bremsstrahlung radiation is a low-efficiency process responsible for x-ray generation. It will occur more in materials with high atomic number. When shielding highly active β sources a two layered shield is used, the inner to shield the β particles, and the outer to shield the x-rays produced from bremsstrahlung radiation.
- γ photons will interact with matter through the photoelectric effect, Compton scatter, or pair production. The photoelectric effect and pair production will remove a photon from the beam, Compton scatter will not.
- Build up describes the additional photons that may arrive at a point of interest after a shield.
- Neutrons may be absorbed or scattered. Generally, they are scattered multiple times, moderating their energy, before being absorbed.

## 10.6    End of Chapter Questions

1. What is the total Bremsstrahlung energy generated by $^{14}$C beta particles?
2. Why is β spectrometry not possible?
3. Hydrogen moderates neutrons with the least number of interactions, why aren't all moderators made of hydrogen?
4. What is the theoretical minimum energy for pair production to occur and why is it not lower?
5. What is the minimum fractional energy after an incident neutron has collided with a atom of $^{207}$Pb, what is it after colliding with an atom of deuterium ($^2$H)?

The hierarchy of controls is a sequential framework that aims to reduce risks. It should be followed sequentially as the safest method, but often the most impractical, is the first step, and the least safe method is the last.

The framework is as follows:

1. Remove the hazard
2. Reduce the hazard
3. Utilise engineering controls
4. Utilise managerial controls
5. Utilise personal protective equipment (PPE)

Engineering controls refer to physical barriers, between an individual and a hazard, that cannot be easily or accidentally overcome or an airflow or ventilation system that removes the hazard. Managerial controls are a set of behavioural practices that ensure the safety of individuals if the individuals follow those practices.

An example of using the hierarchy of controls in radiation protection may be in a calibration facility. Nuclear instrumentation is exposed to various radioactive sources within a calibration facility to ensure it responds as expected. To thoroughly test the instrument, some sources must be sufficiently high in activity to explore the highest detection range and failure mechanisms. These high-activity sources present a significant hazard.

(1) Remove the hazard: Focusing first on the most active sources, unfortunately, this must be practicable. Some poorly designed or damaged instruments can develop faults that cause the instrument to read zero when exposed to a very high dose rate. This is

J. Wain, *Ionising Radiation Protection*, Synthesis Lectures on Engineering, Science, and Technology, https://doi.org/10.1007/978-3-031-65525-8_11

termed *failing to danger*, which may lead an operator to assume a hazardous area is safe. Consequently, the use of these high-activity sources is often vital. However, it may be possible to have two calibration facilities, only one of which contains very high activity sources.

(2) Reduce the hazard: Again, the high-activity sources are vital; however, reducing the activity of some of the other sources used in the calibration process may be possible.

(3) Engineering controls: Engineering controls are going to be of paramount importance. Examples of engineering controls within a calibration facility may include an interlocking key system, which prohibits the sources from being released from their shielding unless the door to the calibration facility is locked, shielding from the walls of the facility to ensure the dose received by the operators is as low as is reasonably practicable, and an installed gamma monitor that prohibits the door to the calibration facility being opened if a high dose rate is detected.

(4) Managerial controls: Managerial controls will include any workplace practice that ensures the safety of individuals. For instance, only fully trained operators can operate the calibration facility; operators must use the installed CCTV system to ensure no one has been inadvertently locked in the calibration room before a source is exposed.

(5) Personal protective equipment (PPE) has limited uses within a calibration facility. Almost all sources used within a calibration facility are certified to have a certain level of physical robustness, so PPE will not be required to inhibit radioisotopes from entering the body. Lead vests are sometimes used in other industries for protection against low-energy photons; however, they offer very little protection against high-energy photons, so they are not used in calibration facilities.

Methods for ionising radiation protection may also be considered by focusing on the hazard and classifying it based on principles explained in the previous chapters.

External hazards: External hazards will be from gamma, x-ray, and neutron sources. They should be mitigated by reducing the time, increasing the distance, and increasing the shielding.

Internal hazards: mitigate ingress pathways: ingestion, inhalation, injection, absorption.

## 11.1  Principles of Protection ICRP

The ICRP lists three main principles of protection: justification, optimisation, and limitation.

Justification refers to the use of radiation in a manner where the benefits outweigh the disadvantages. If there is another way of achieving the same outcome without the use of radioactive substances, that method should be pursued.

Optimisation refers to the amount of radiation; the source strength must be sufficient to achieve the necessary result and not greater than required. Optimisation is closely

**Table 11.1**  Annual dose limits as stipulated by the Ionising Radiation Regulations 2017 (IRRs 17)

|  | Whole body effective dose (mSv) | Skin and extremities (mSv) | Lens of the eye (mSv) |
|---|---|---|---|
| Any employee or trainee (>18 yrs old) | 20 | 500 | 20 |
| Trainee (<18 yrs old) | 6 | 150 | 15 |
| Any other person | 1 | 50 | 15 |

related to the principle of ALARP, which states that doses should be kept As Low As Is Reasonably Practicable.

Limitation refers to the need to adhere to absolute dose limits. In the UK, these are set out in the Ionising Radiation Regulations 2017 (IRRs 17) (Table 11.1).

'Any other person' includes members of the public. This group has the most conservative dose limits as they may be unaware of ongoing exposure, not have enough information to understand its consequences, and will not receive any personal benefit, unlike a radiation worker. The dose limits for the skin are applied to the dose averaged over any area of 1 $cm^2$, regardless of the area exposed. The dose limits to the skin, extremities, and eye lens ensure exposed individuals do not exceed the threshold for harmful tissue reactions.

## 11.2  Radiation Workers

Radiation workers may be exposed to a greater level of hazard. They thus must be able to make an informed decision as to whether to accept this additional risk, and further risks must be mitigated. Within the IRRs, radiation workers may be designated as *classified persons* if their potential risk deems it necessary and they are over 18 years old. This potential risk is defined as radiation workers who may be exposed to an effective dose >6 mSv/y, an equivalent dose >15 mSv/y for the lens of the eye, or >150 mSv/y for the skin or extremities. Workers may also need to become classified persons if they work with sources capable of giving a high dose rate that may result in an employee receiving >20 mSv or exceeding any other dose limit within a few minutes.

Before being designated as a classified person, radiation workers must undergo a medical examination with a relevant doctor. This doctor must consider the fitness of the worker to wear any required PPE, any skin conditions that may facilitate the ingress of radioisotopes through the skin, psychiatric illnesses or personality disorders that may impact the ability of an individual to take responsibility for safety, and any histories of chronic pulmonary disease, blood disorders, treatments with cytotoxic drugs, genetic predispositions to cancers, or previous significant medical exposures to radiation. Classified persons will also undergo ongoing medical surveillance to ensure they remain fit to be classified persons.

Classified persons must also undergo additional training commensurate with the type and extent of hazard they are reasonably foreseeable to encounter. This training will cover the basics of radiation protection, the behaviours required to reduce external and internal exposures, and the contingency plans to enact to circumvent or mitigate radiation accidents.

After designating an individual as a classified person, the employer must be able to record and store the dose received by the individual. This esoteric capability is typically outsourced to an Approved Dosimetry Service (ADS).

In addition to classified persons, non-classified persons, such as radiation workers, do not meet the requirements to become classified persons. The requirements for non-classified persons are loosely similar to those of classified persons and are matched to their level of risk from ionising radiation.

## 11.3   Areas Containing a Radiation Hazard

Some areas will have a higher hazard and thus must have additional controls. These areas may be designated as *controlled* or *supervised* areas. A controlled area is one in which special procedures must be followed to limit significant exposure or to mitigate the impact of radiation accidents. Or an individual working in the area may be subjected to an effective dose exceeding 6 mSv/y, or an equivalent dose exceeding 15 mSv annually for the eye lens, or 150 mSv annually for the skin or extremities. If an area is designated as controlled, the employer must ensure they can control the individuals who enter it. Ideally, this will be achieved with a physical barrier, i.e. a locked door with keys only accessible to the correctly designated individuals who need access to that area. These individuals will be classified, or non-classified persons operating under a specific risk assessment termed a *written arrangement*.

Supervised areas will be designated according to the dose rates within each area and the likelihood that these rates may change. A supervised area will be required if an individual working there will likely receive over 1 mSv/y.

These areas will require frequent surveys to determine if the dose rates have changed. The frequency of these surveys will correspond to the level of risk.

## 11.4   Summary

- The hierarchy of controls is a sequential process to mitigate hazards.
- The principles of ionising radiation protection are justification, optimisation, and limitation.

- Radiation workers exposed to greater levels of risk will be designed as classified persons. They must be older than 18, have passed a medical examination, and be specially trained in radiation protection and relevant contingency plans.
- Areas that contain a radioactive hazard may be designated as supervised or controlled areas. Supervised areas will be associated with a whole-body effective dose of between 1 and 6 mSv/y or an area that must be reviewed. Controlled areas will be related to a whole-body effective dose of >6 mSv/y or require special precautions to limit significant doses.

## 11.5  Exercises

1. List the hierarchy of controls.
2. What is the dose limit for the lens of the eye for a 17-year-old apprentice?
3. When will a radiation worker be required to be designated as a classified person?
4. When will an area be required to be a supervised area?
5. Contamination has been leaking into a supervised area from the floor above. A recent survey showed an elevated dose rate of 15 $\mu$Sv/h. Should the designation of this room change?

# Nuclear Emergencies

The UK's Radiation Preparedness and Public Information Regulations 2019 (REPPIR) define a radiation emergency as 'a non-routine situation or event arising from work with ionising radiation that necessities prompt action to mitigate serious consequences…to human life, health and safety, quality of life, property, or the environment.' The approved code of practice for REPPIR defines serious consequences as a dose to a member of the public exceeding 1 mSv in the year following the event. For that to occur, a substantial release must occur on the site, and any containment must be severely compromised. A radiation emergency is a major event.

Radiation emergencies can be divided into two categories: those resulting from malicious or non-malicious events. While there is significant overlap, this chapter will focus on non-malicious events.

## 12.1  Nuclear Reactor Overview

Most power-producing nuclear reactors worldwide are pressurised water reactors (PWRs) or boiling water reactors (BWRs). There are many different nuanced designs of PWRs and BWRs; for brevity the nuances will be omitted and each reactor type will be described in general.

Both reactor designs have a core containing uranium oxide, enriched to 3–5% in $^{235}$U. This uranium is in the form of small cylindrical pellets, sometimes with a hollow core, stacked upon one another and placed within a metal tube known as cladding. This tube is sealed at both ends to contain the uranium and any fission products. These tubes are arranged into fuel assemblies and placed into the reactor's core. When the reactor is

© The Author(s), under exclusive license to Springer Nature Switzerland AG 2025    81
J. Wain, *Ionising Radiation Protection*, Synthesis Lectures on Engineering, Science, and Technology, https://doi.org/10.1007/978-3-031-65525-8_12

critical, neutrons are captured by the $^{235}$U atoms, resulting in the $^{235}$U undergoing fission, producing two fission products, more neutrons, and heat. The term *critical* describes a stable state where the nuclear chain reaction is self-sustaining, with each fission producing enough neutrons to create another fission. Additionally, heat is produced from the fission products as they radioactively decay; this is why it is essential to maintain core cooling for a time after the reactor is no longer critical. The heat generated within the core is used to heat water that is circulated throughout the reactor.

A PWR has two coolant circuits, a primary and a secondary circuit. The primary circuit is held at a high pressure, enabling the water in this circuit to reach temperatures exceeding 300 °C without boiling. A steam generator links this circuit to the secondary circuit where steam is generated, which turns steam turbines, spins a generator, and generates electricity. The water in this secondary circuit is cooled by a large body of water via a heat exchanger before being recirculated to the steam generators.

A BWR only has one coolant circuit, and the water is directly boiled within the core, producing the steam that is fed into the turbines. Again, this water must be cooled by a large body of water before being fed back into the core.

In PWRs, BWRs, and all other reactor designs, many layers of containment must be breached before significant radioactivity is released. In a PWR, one such pathway might be fission products escaping the chemical matrix of the uranium oxide fuel, breaching the cladding, entering the primary coolant circuit, breaching the primary coolant circuit to enter the containment building, and finally, breaching the containment building to be released into the atmosphere.

Modern nuclear reactors are exceptionally well designed, employing many methods of circumventing or mitigating accident scenarios. This is reflected by the rarity of nuclear emergencies and the root causes of the few that have occurred. PWRs and BWRs are intrinsically stable designs. They use water to both remove the heat from the core and moderate the neutrons. Consequently, if the reactor loses coolant, it will also lose the ability to slow the neutrons down and the fission chain reaction will stop.

A typical nuclear emergency follows some generic steps. First, the reactor will lose the ability to remove heat from the core. This heat will gradually build up, the cladding will melt or warp, fission products will be released into the cooling circuit, and eventually escape into the atmosphere.

The loss of the ability to remove heat from a reactor can occur due to a loss of circulation, a loss of coolant, or the loss of the ultimate heat sink: the large body of water that cools the coolant after the steam generators. Modern reactors have various emergency-core cooling systems to mitigate these scenarios, often including battery back-ups, diesel generators, or passive convection systems.

If all these systems fail, the hazards from a nuclear emergency will take the form of direct γ shine, cloud shine, ground shine, inhalation, and ingestion. Direct γ shine refers to γ radiation emanating from the facility. This hazard will reduce as the distance increases in accordance with the inverse square law. Thus, if you double your distance from the source you will quarter the dose rate. Consequently, this dose pathway becomes inconsequential for all but the closest members of the public. Cloud shine describes the dose received from γ emitters released as a plume from the reactor. This plume, just like smoke from a fire, will travel downwind. The plume will contain γ emitters that will irradiate individuals from above. Cloud shine will only be a relevant dose pathway when the plume is nearby. A more concentrated plume will be more hazardous, the further the plume travels from its sources the more dilute it will become. The rate of plume dilution will be dependent upon the specific weather conditions during the release. The average windspeed in the UK is around 10 mph (16 kph), about 5 m/s; this feels like a gentle breeze. At this windspeed, a plume will take a little over three minutes to travel a kilometre. Consequently, after a release has stopped, the plume can be quickly diluted, reducing the cloud shine and inhalation hazard. Radionuclides from the plume will deposit on the ground and irradiate individuals from below; this dose pathway is termed ground shine. Ingestion refers to eating or drinking material that has been radioactively contaminated. This occurs when material from the plume falls onto crops, pasture, and water sources and enters human food chains.

## 12.2 How to Respond

In a nuclear emergency, the best thing to do is follow government advice. Every nuclear facility that contains significant quantities of radioactive material will have undertaken a hazard evaluation and consequence assessment. This is a legal stipulation under REPPIR. Fundamentally, it is a detailed risk assessment that identifies potential emergency scenarios and ascertains the result of each and their consequences. These emergency scenarios are compiled alongside their likelihoods. After this has been completed, a generic plan for responding to a nuclear emergency is created. This generic plan will be designed to mitigate against all but the most severe and unlikely emergency scenarios, and if one of these does occur, the plans are designed to enable them to be extended. It takes years of work from teams of experts to create these emergency response plans. The government advice issued following a nuclear emergency will be based on this work.

There are only four main actions to be taken following a nuclear emergency: do not ingest anything that may have been exposed to radioactive material, take stable iodine tablets, shelter, and evacuate. Every mitigation will have advantages and disadvantages; government advice will balance these before issuing recommendations. Following a nuclear emergency, food and water restrictions are likely to be widespread. These will be so widespread because we live in a society where it is easy to acquire food and water from other parts of the country or even the world, so the advantages of avoiding even a tiny dose will outweigh the disadvantages of altering the supply chain.

Stable iodine tablets mitigate the hazard of radioactive iodine. Our bodies will uptake iodine and concentrate it in the thyroid, a small organ near your spine at the base of your neck. Stable iodine saturates the thyroid with non-radioactive iodine, inhibiting the body's uptake of radioactive iodine. There are several isotopes of iodine produced as fission products, and the thyroid dose is often one of the main contributors to the dose to members of the public following a nuclear emergency; however, iodine radioisotopes all have short half-lives. The most significant isotope is $^{131}$I, with a half-life of around eight days. Hence, iodine will only be a hazard if the nuclear emergency involves a nuclear power plant that has been recently operational.

Shelter involves staying within a building, closing windows and shutting off ventilation systems. Sheltering can reduce the dose from inhalation by around 50% and the cloudshine dose by around 80%, depending upon the building in which an individual is sheltering.

Evacuation is removing individuals from the immediate vicinity and downwind of the release. Evacuation has the most significant disadvantages of any response measure. In every recent nuclear emergency, fatal car crashes have occurred after evacuation advice has been given. The dose that must be avoided to justify evacuating a large number of people must be high. If not, individuals will risk their lives escaping from a potentially inconsequential hazard. Additionally, if an evacuation is poorly timed, individuals may leave the relative safety of their homes to be held up in a traffic jam as a plume passes overhead. Cars offer minimal reduction in the cloud shine dose.

## 12.3    Nuclear Emergencies in Perspective

Intrinsically, a radiation emergency can be a terrifying prospect for members of the public. Being harmed by something that you cannot sense and that you may not fully understand can have tremendous psychological health effects. It is often these psychological effects that are the most harmful. The worst nuclear emergency the world has experienced occurred in 1986 at the Chornobyl nuclear power plant when an intrinsically flawed reactor design was operated poorly. The effects of this disaster were far-ranging and persist in the public imagination. Chornobyl is often still used as an example by antinuclear organisations in an attempt to deter progress in the field of nuclear electricity generation. However, the total death toll from Chornobyl across all time is stated by the World Health Organisation, International Atomic Energy Agency, and the United Nations as less than 4,000 [13]. Other estimates abound, and the author advises the astute reader to consider the motives of the organisations that publish them. For comparison, road traffic incidents in the US in 2021 killed 42,939 [14], the flu season in the UK from 2022 to 2023 killed 14,500 people [15], and diabetes in Australia killed 5,404 people in 2023 [16]. Every death is an individual tragedy, yet the public perception of the risks of the nuclear industry is greatly exaggerated.

## 12.4   A Chronological Summary of Major Nuclear Emergencies

**Kyshtym Disaster**—September 29, 1957

Summary: An explosion at the Mayak Production Association in the Soviet Union led to the release of radioactive materials. It is considered one of the worst nuclear accidents in history.

INES Scale: The INES scale was not used at the time, but retrospectively, it would likely be classified as a Level 6 or 7 event.

**Windscale Fire (Sellafield)**—October 10, 1957

Summary: A fire broke out at the graphite-moderated air-cooled Windscale nuclear reactor in the United Kingdom, releasing radioactive gases into the atmosphere.

INES Scale: Retrospectively, it would likely be classified as a Level 5 event.

**SL-1 Reactor Accident**—January 3, 1961

Summary: A criticality excursion occurred at the SL-1 experimental reactor in Idaho, USA, resulting in an explosion that caused the deaths of three personnel and significant radioactive contamination.

INES Scale: Retrospectively, it would likely be classified as a Level 4 or 5 event.

**Three Mile Island Accident**—March 28, 1979

Summary: A partial core melt occurred at the Three Mile Island Nuclear Generating Station in Pennsylvania, USA. The operators isolated an emergency core cooling system, started the plant against the operating guidance, and responded poorly after a valve became stuck open, resulting in a slow but sustained release of coolant water. Despite this, there was no significant radiation hazard.

INES Scale: Level 5 (Accident with Wider Consequences).

**Chornobyl Nuclear Disaster**—April 26, 1986

Summary: An explosion and fire at the graphite-moderated water-cooled Chornobyl Nuclear Power Plant in Ukraine released a significant amount of radioactive material into the atmosphere. 28 people died because of acute radiation syndrome within a few weeks of the incident; the total number of deaths is estimated as $< 4,000$. The accident occurred because of a flawed reactor design and serious mistakes made by the operators.

The Chornobyl Power Complex contained four RBMK-1000 reactors. Units one, two, and three continued operating until they were shut down in 1997, 1991, and 2000, respectively. Following the disaster, around 6,000 personnel worked at these remaining units.

INES Scale: Level 7 (Major Accident).

**Goiania Radiological Accident**—September 13, 1987

Summary: In Brazil, a radioactive cesium-137 source was inadvertently scavenged from an abandoned hospital by individuals hoping to sell scrap metal, leading to widespread contamination and radiation exposure in the local population.

INES Scale: Level 5 (Accident with Wider Consequences).

**Fukushima Daiichi Nuclear Disaster**—March 11, 2011

Summary: A massive earthquake and tsunami in Japan led to the meltdown of three reactors at the Fukushima Daiichi Nuclear Power Plant. It resulted in the release of radioactive materials and was the most severe nuclear accident since Chornobyl.

INES Scale: Level 7 (Major Accident).

## 12.5  Summary

- In the Radiation Emergency Preparedness and Public Information Regulations of 2019, the definition of a nuclear emergency includes members of the public receiving >1 mSv in the year following the incident.
- Nuclear emergencies typically occur when the core of a nuclear reactor gets too hot, and this excess of heat damages the fuel.
- There are five dose pathways following a nuclear emergency: direct $\gamma$ shine, cloudshine, groundshine, inhalation, and ingestion.
- Following a nuclear emergency, there are four methods for limiting an individual's dose: do not ingest anything that may have been exposed to the radioactive plume, take stable iodine tablets, shelter, and evacuate.
- Each countermeasure has advantages and disadvantages; evacuation has the most significant disadvantages and widespread evacuations typically result in fatalities.
- The World Health Organisation, International Atomic Energy Agency, and the United Nations state the total number of deaths from Chornobyl is less than 4,000.

## 12.6  Exercises

1. A nuclear emergency has occurred at a site that manufactures fuel for nuclear power plants. How effective will stable iodine tablets be?
2. What response would you take to mitigate a cloud shine dose?
3. What is the main difference between PWRs and BWRs?
4. What makes PWRs and BWRs intrinsically safe?
5. Why is direct $\gamma$ shine not a concern over extended distances?

# Solutions to End of Chapter Exercises

## Introductory Nuclear Physics

1. What principle is commonly used to separate U-238 from U-235?
   The different masses. As the isotopes are both uranium, the chemistry will be identical.
2. What is the theoretical minimum energy required for pair production?
   The rest of the mass of the two electrons is 1.022 MeV.
   This may be calculated by:

$$E = mc^2$$

The rest mass of a single electron is $9.1094 \times 10^{-31}$ kg
The speed of light in a vacuum is $2.998 \times 10^8$ ms$^{-1}$
There are $6.242 \times 10^{18}$ eV in a J and there are two electrons being produced by pair production.

$$E = 9.1094 \times 10^{-31} \times \left(2.998 \times 10^8\right)^2 \times 6.242 \times 10^{18} \times 2$$
$$= 1.022 \, \text{MeV}$$

3. What is the decay constant for Cs-137 with a half-life of 30 years?
   $\lambda = \ln 2/t_{1/2}$
   30 years $= 9.467 \times 10^8$ s
   $\lambda = 732.17 \times 10^{-12}$ s$^{-1}$
4. How long will it take for 95% of a sample of Cs-137 to decay?
   $N_t = N_0 e^{-\lambda t}$
   $5/95 = e^{-\lambda t}$
   $n\,(5/95)/-\lambda = t$
   $4.021 \times 10^9$ s $= 127$ y

© The Editor(s) (if applicable) and The Author(s), under exclusive license                     87
to Springer Nature Switzerland AG 2025
J. Wain, *Ionising Radiation Protection*, Synthesis Lectures on Engineering, Science, and
Technology, https://doi.org/10.1007/978-3-031-65525-8

5. What is the difference between X-rays and gamma radiation?
   How they are produced. Gamma radiation is produced from an atomic nucleus, and
   X-rays are produced from processes outside an atomic nucleus.

## Key Units

1. What mass of $^{238}$U would be required for a 1 GBq source?

$$m = AM \, / \, \lambda N_A$$

   80.9 kg
2. What mass of $^{60}$Co would be required for a 1 GBq source?
   $2.39 \times 10^{-5}$g
3. What is the absorbed dose from a 14 J exposure to a 70 kg adult?
   $14/70 = 0.2$ J/kg $= 0.2$ Gy
4. What will be the whole-body effective dose if an individual receives a 3 Gy absorbed
   dose to their entire body from a gamma-emitting radioisotope?
   $3 \times 1 \times 1 = 3$ Sv
5. What will be the whole-body effective dose if an individual receives a 30 mGy
   exposure to their lungs from an alpha-emitting radioisotope?
   $30 \times 10^{-3} \times 0.12 \times 20 = 0.072$ Sv or 72 mSv

## Cellular and Whole-Body Responses to Radiation

1. What is the most radiosensitive part of the cell?
   The nucleus
2. Describe indirect radiation damage
   A chemically reactive species is created from ionising radiation. This species then
   damages the cell.
3. Why does the prevalence of harmful tissue reactions plateau at higher doses?
   There is a finite number of cells or amount of tissue that can be damaged.
4. What is the lifespan of a white blood cell?
   1–3 days
5. What will be the increased risk of developing cancer to a member of the public after
   receiving a chronic dose of 20 mSv?
   $20 \times 10^{-3} \times 5 = 0.1\%$

## Epidemiological Studies and Radiation Risk

1. List some difficulties with epidemiological studies that focus on low-dose chronic exposure.
   Finding an equivalent population with the same cancer risk factors, a large enough study group, and statistically significant variations in cancer prevalence in the exposed group.
2. What were the main conclusions from the Radon and Miners studies?
   There is a clear link between high levels of radon gas exposure and an increased risk of lung cancer.
3. What is the average background dose in the UK? 2.7 mSv/y
4. What is the increased cancer detriment to a worker from an exposure of 100 mSv?

$$100 \times 10^{-3} \times 4.1 = 0.41\%$$

5. Why does the lens of the eye have a specific dose limit?
   This is to protect the lens of the eye from cataracts. Cataracts are a type of harmful tissue reaction, so they have a threshold value.

## Internal Radiation Exposure

1. The radioactive half-life of tritium is 12.35 years, the biological half-life is ten days, what is the effective half-life?
   9.97 days
2. How long will it take for 90% of a sample of tritium to decay?
   41 years
3. How long will it take for 90% of tritium to leave the body after an intake?
   33.1 days
4. What is the CED to a worker after ingesting 10 MBq of Cs-137?
   130 mSv
5. What is the mass of Pu-238, in oxide form, that will result in a CED of 5 Sv if inhaled as 1 $\mu$m particles?
   $m = (\text{CED}/\text{DPUI}) \times (1/\lambda) \times (M/N_a)$
   $m = 5/5 \times 10^{-5} \times (2.769 \times 10^9/\ln 2) \times 238/6.022 \times 10^{23}$
   $= 1.58 \times 10^{-7}$ g

## External Radiation Exposure

1. How much dose will be received after spending 20 min in a dose rate of 80 mSv/h?
   $(20/60) \times 80\,\text{mSv} = 26.7\,\text{mSv}$

2. How much dose will be received after spending 8 s in a dose rate of 6.4 Sv/h? $(6.4/60 \times 60) \times 8 = 14.2 \, mSv$

3. The dose rate is 4 mSv/h at 2.3 m; what is the dose rate at 9 m?

$$4 \times 10^{-3} \times 2.3^2/9^2 = 0.26 \, mSv/h$$

4. The dose rate is 2 Sv/h at 1.4 m; at what distance will the dose rate be 100 mSv/h?

$$x_2 = \sqrt{D_1 . x_1^2 \big/ D_2}$$

6.26 m

5. The dose rate is 1.6 Sv/h at 4.2 m; at what distance will you accue a dose of 200 mSv in 15 min?

200 mSv in 15 min is equivalent to 800 mSv/h $((60/15) \times 200)$

$$x_2 = \sqrt{D_1 . x_1^2 \big/ D_2}$$

5.94 m

---

## Background Radiation Exposure

1. What are the three categories of background radiation?
   Primordial, cosmic, anthropogenic

2. Why is there so much focus on radon?
   Radon exists as an alpha-emitting gas, so it has a high $W_R$, high mobility, and the ability to be trapped and accumulate at low points in dwellings; about 50% of all background radiation dose in the UK can be attributed to exposure from radon gas, and its concentration can vary widely depending on the bedrock of the area and human behaviours.

3. What was the isotopic concentration of $^{235}$U at the time of the Oklo reactors operation?

$$A_t = A_0 e^{-\lambda t}$$

$$A_0 = A_t \big/ e^{-\lambda t}$$

The half-life of $^{235}$U is 700 million years, therefore $\lambda = \ln 2/700 = 9.90 \times 10^{-4}/$ million y

The Oklo reactors were undergoing fission around 1.7 billion years ago.

The concentration of $^{235}$U today is 0.72%.

Assuming the concentration of the other uranium isotopes has not changed significantly, 1.7 billion years ago 3.88% of natural uranium was $^{235}$U.
4. Why do aircrews experience increased levels of background radiation?
They spend longer at high altitudes and are, therefore, less shielded by the atmosphere from cosmic background radiation.
5. What percentage of background radiation in the UK is from medical exposures?
16%

## Radioactive Waste Management

1. Granite rock often contains uranium ore, so is it a radioactive waste?
No
2. What is the key difference between HLW and ILW?
HLW is heat-generating.
3. What is the main advantage of the delay and decay principle of radioactive waste management?
The radioactive material is no longer radioactive and will not constitute a hazard.
4. Why will used fuel remain hazardous for a long time?
Used fuel consists of many isotopes, from fission products to activation products and neutron capture reactions. Some of these isotopes have very long half-lives.
5. How might you dispose of $^{131}$I?
$^{131}$I has a half-life of about eight days, so it is well suited to the delay and decay principle.

## Radiation Detection and Instrumentation

1. What energy band gap will create visible light?

$$E = hc/\lambda$$

The wavelength of visible light ranges from 380 to 700 nm.
$1\,J = 6.242 \times 10^{18}$ eV
$h = 6.626 \times 10^{-34}$ Js
$c = 2.998 \times 10^{8}$ ms$^{-1}$
between $2.8 \times 10^{-19}$ J and $5.2 \times 10^{-19}$ J
or 1.7 eV and 3.2 eV
2. What are the main advantages of semiconductor detectors?
A very fast response time and a low ionisation energy that creates more electrons per joule deposited in the detector and enables significantly better resolution.

3. What is the main advantage of a GM tube?

   The Townsend avalanches and additional ionisations created from UV photons act as an intrinsic amplification to the output pulse.

4. List five advantages of TLDs.

   They are a passive form of detection requiring no batteries or user input. They are cheap, robust, fairly accurate, have a large range of energy sensitivities, are lightweight and easy to wear.

5. How many steradians makeup half a sphere?

   $2\pi$ steradians

## Shielding

1. What is the total Bremsstrahlung energy generated by $^{14}$C beta particles?

   0.02 keV

2. Why is $\beta$ spectrometry not possible?

   Spectrometry relies on matching the energy of a particle or photon to a specific isotope. $\beta$ particles share their energy with an antineutrino so are produced with a spectrum of energies.

3. Hydrogen moderates neutrons with the least number of interactions, why aren't all moderators made of hydrogen?

   The density and engineering challenges of the material must also be considered. Hydrogen gas is not dense and explosive, so hydrogen moderators typically take the form of plastics or water. Hydrogen will also absorb more neutrons than other moderators.

4. What is the theoretical minimum energy for pair production to occur and why is it not lower?

   1.022 MeV this is the rest mass of an electron

5. What is the minimum fractional energy after an incident neutron has collided with a atom of $^{207}$ Pb, what is it after colliding with an atom of deuterium ($^2$H)?

   $\alpha = (A - 1/A + 1)^2$

   Pb = 98%

   H-2 = 11%

## Operational Radiation Protection

1. List the hierarchy of controls.

   Remove the hazard, reduce the hazard, engineering controls, managerial controls, and PPE.

2. What is the dose limit for the lens of the eye for a 17-year-old apprentice?

15 mSv/y

3. When will a radiation worker be required to be designated as a classified person?

   If their dose may exceed 6 mSv/y, an equivalent dose of 15 mSv/y for the lens of the eye, or 150 mSv/y for the skin or extremities or if they are working with sources that may exceed a dose limit in several minutes.

4. When will an area be required to be a supervised area?

   If the area must be kept under review, or if the dose rates in the area result in an annual dose between 1 and 6 mSv.

5. Contamination has been leaking into a supervised area from the floor above. A recent survey showed an elevated dose rate of 15 μSv/h. Should the designation of this room change?

   This will depend upon the room's occupancy. Assuming a worker occupies the room for 1,000 hrs a year, a dose rate of 15 μSv/h will result in 15 mSv/y. Therefore, the room should be designated as a controlled area.

## Nuclear Emergencies

1. A nuclear emergency has occurred at a site that manufactures fuel for nuclear power plants. How effective will stable iodine tablets be?

   Not at all effective. Stable iodine tablets reduce the dose received from radioactive iodine. Radioactive iodine is produced as a fission product. A site that manufactures new nuclear fuel will not have any fission products.

2. What response would you take to mitigate a cloudshine dose?

   Shelter or evacuation

3. What is the main difference between PWRs and BWRs?

   PWRs have two coolant loops, a pressurised primary loop and a secondary loop, with a steam generator enabling heat exchange between them.

4. What makes PWRs and BWRs intrinsically safe?

   They both use water as a moderator and a coolant, so if there is no coolant, there is also no moderator, and the fissions will stop.

5. Why is direct γ shine not a concern over extended distances?

   The γ radiation will be geometrically reduced following an inverse square relationship.

# Key Organisations

## International Commission on Radiological Protection (ICRP)

The ICRP is an independent international organization that provides recommendations and guidance on radiation protection. It was established in 1928 and is headquartered in Ottawa, Canada. The primary mission of the ICRP is to develop and disseminate recommendations and guidance on the protection of people from the harmful effects of ionizing radiation.

The ICRP regularly updates its recommendations to reflect advances in scientific understanding and technology. Its work covers a wide range of applications, including medical radiation, industrial uses of radiation, and protection of the public and the environment.

The recommendations produced from the ICRP are intended to create a set of basic principles and indicate the consequences of radiation exposure. However, 'the estimation of these consequences and their implications necessarily involves social and economic judgements as well as scientific judgements in a wide range of disciplines'.[1] These principles are not legally binding and are instead used on the national scale to create detailed legislation, recommendations, and codes of practice. In the UK, the ICRP recommendations form the foundations for the Ionisation Radiation Regulations.

## International Atomic Energy Agency (IAEA)

The International Atomic Energy Agency (IAEA) is an international organization, goverened by its member states, that promotes the peaceful use of nuclear energy and works to prevent its use for any military purpose, including nuclear weapons. Its General Conference, consisting of all member states, meets annually to set policies and approve the budget. The organization is headquartered in Vienna, Austria. The IAEA was established in 1957 as an autonomous organization under the United Nations (UN) in response to the "Atoms for Peace" initiative proposed by U.S. President Dwight D. Eisenhower. Key functions and responsibilities of the IAEA include:

© The Editor(s) (if applicable) and The Author(s), under exclusive license
to Springer Nature Switzerland AG 2025
J. Wain, *Ionising Radiation Protection*, Synthesis Lectures on Engineering, Science, and Technology, https://doi.org/10.1007/978-3-031-65525-8

*Promoting the Peaceful Use of Nuclear Energy*
The IAEA supports the development and application of nuclear energy for peaceful purposes, such as electricity generation, medical uses, and industrial applications.

*Safeguards and Verification*
The agency plays a crucial role in verifying that countries are using nuclear materials for peaceful purposes and not for the development of nuclear weapons. It implements safeguards agreements to ensure compliance with the Treaty on the Non-Proliferation of Nuclear Weapons (NPT).

*Safety Standards*
The IAEA establishes international safety standards for nuclear power plants and other nuclear facilities to ensure the protection of people and the environment from the harmful effects of radiation.

*Technical Assistance*
The agency provides technical assistance and cooperation to its member states, particularly developing countries, in areas such as nuclear energy, radiation protection, and nuclear safety.

*Nuclear Security*
The IAEA works to enhance global nuclear security by assisting member states in securing nuclear materials and facilities to prevent unauthorized access or malicious use.

*Research and Development*
The IAEA conducts research and development activities in various areas related to nuclear science and technology.

## International Commission on Radiological Units and Measurements (ICRU)

ICRU is a non-profit organization that was established in 1925 and now works closely with the ICRP. The primary purpose of the ICRU is to develop internationally accepted recommendations and standards for the use of ionizing radiation in medicine and industry, as well as for radiation protection. Key activities of the ICRU include:

*Standardization of Units*
The ICRU is involved in the standardization of quantities, units, and nomenclature used in the measurement of ionizing radiation.

*Dosimetry*

The ICRU provides guidance on dosimetry and provides recommendations for practical measurements that result in uniformity of reporting.

# Terminology

Activity—the number of nuclear decays per second, measured in Becquerels (Bq), it relates to the potency of a radioactive source.

Anthropogenic—relating to or caused by human activity.

Bremsstrahlung—the electromagnetic radiation released by a charged particle when slowed down by another charged particle.

DNA—Deoxyribonucleic Acid, an organic polymer comprised of two polynucleotide chains, linked by base pairs, that coil around one another to form a double helix. DNA carries genetic information and enables cell reproduction.

Dose, Absorbed—the amount of energy deposited per kilogram, measured in Gy.

Dose, Effective—the amount of energy deposited per kilogram multiplied by a radiation weighting factor and multiplied by a tissue weighting factor, measured in Sv.

Dose, Equivalent—the amount of energy deposited per kilogram multiplied by a radiation weighting factor, measured in Sv.

Half Value Layer—a concept used in shielding that dictates the thickness of a material required to reduce a radiation field by half.

Half-Life—the time required for half of the radioactive isotopes to undergo a nuclear decay.

Harmful Tissue Reactions—biological responses to a radiation dose that will occur after a specific dose threshold has been exceeded, sometimes referred to as deterministic effects.

Ionisation—the removal of an orbital electron, resulting in a free electron and a positively charged ion.

Isotope—a chemical species with the same number of protons, and thus is the same element, but a different number of neutrons.

Neutron—a sub-atomic particle with no charge.

Photon—a packet of electromagnetic energy.

Stochastic—refers to a process that exhibits a probability and can be analysed statistically but cannot be predicted precisely.

J. Wain, *Ionising Radiation Protection*, Synthesis Lectures on Engineering, Science, and Technology, https://doi.org/10.1007/978-3-031-65525-8

# Scientific Notation

| Tera | T | $10^{12}$ |
|------|---|-----------|
| Giga | G | $10^{9}$ |
| Mega | M | $10^{6}$ |
| Kilo | k | $10^{3}$ |
| Hecto | h | $10^{2}$ |
| Deca | da | 10 |
| Deci | d | 0.1 |
| Centi | c | 0.01 |
| Milli | m | 0.001 |
| Micro | $\mu$ | $10^{-6}$ |
| Nano | n | $10^{-9}$ |

© The Editor(s) (if applicable) and The Author(s), under exclusive license
to Springer Nature Switzerland AG 2025
J. Wain, *Ionising Radiation Protection*, Synthesis Lectures on Engineering, Science, and
Technology, https://doi.org/10.1007/978-3-031-65525-8

# Half-Lives of Common Or Useful Radioisotopes

Potassium-40 (K-40)—~1.25 billion years (used in geological dating).

Cobalt-60 (Co-60)—~5.27 years (used in radiation therapy and industrial radiography).

Krypton-85 (Kr-85)—10.8 years (used in leak detection and industrial applications)

Technetium-99 (Tc-99)—211,100 years (used in radiopharmaceuticals).

Technetium-99m (Tc-99m)—6 h (used in medical imaging).

Iodine-123 (I-123)—13 h (used in medical imaging).

Iodine-131 (I-131)—~8 days (used in nuclear medicine for thyroid treatment).

Cesium-137 (Cs-137)—~30 years (used in radiation therapy and industrial applications).

Uranium-235 (U-235)—~703.8 million years (used in nuclear reactors and weapons).

Uranium-238 (U-238)—~4.468 billion years (used in geological dating).

J. Wain, *Ionising Radiation Protection*, Synthesis Lectures on Engineering, Science, and
Technology, https://doi.org/10.1007/978-3-031-65525-8

# Comparative Whole-Body Effective Doses

Comparative whole-body effective doses in mSv

| | |
|---|---|
| Dental X-ray | 0.005 |
| 100g of Brazil nuts | 0.01 |
| Chest X-ray | 0.014 |
| Transatlantic flight | 0.08 |
| Nuclear power station worker average annual occupational exposure (2010) | 0.18 |
| CT scan of the head | 1.4 |
| UK average annual radiation dose | 2.7 |
| USA average annual radiation dose | 6.2 |
| CT scan of the chest | 6.6 |
| Average annual radon dose to people in Cornwall | 7.8 |
| CT scan of the whole spine | 10 |
| Annual exposure limit for UK nuclear industry employees | 20 |
| Observable changes in blood cells | 100 |
| Acute radiation effects including nausea and a reduction in white blood cell count | 1000 |
| Dose of radiation which would kill about half of those receiving it in a month | 5000 |

© The Editor(s) (if applicable) and The Author(s), under exclusive license
to Springer Nature Switzerland AG 2025
J. Wain, *Ionising Radiation Protection*, Synthesis Lectures on Engineering, Science, and
Technology, https://doi.org/10.1007/978-3-031-65525-8

# Further Reading

The elements of nuclear power—D J Bennet.

Introductory Nuclear Physics—Kenneth S Krane.

Radiation Detection and Measurement—Glenn F Knoll.

J. Wain, *Ionising Radiation Protection*, Synthesis Lectures on Engineering, Science, and
Technology, https://doi.org/10.1007/978-3-031-65525-8

# Bibliography

1. Vennart, J. The 1990 recommendations of the International Commission on Radiological Protection. *J. Radiol. Prot.* **11**, 199–203 (1991).
2. Life Span Study (LSS) – Radiation Effects Research Foundation (RERF). https://www.rerf.or.jp/en/programs/research_activities_e/outline_e/proglss-en/.
3. The Ionising Radiations Regulations 2017. https://www.legislation.gov.uk/uksi/2017/1075/contents/made.
4. Eckerman, K., Harrison, J., Menzel, H.-G. & Clement, C. H. ICRP Publication 119: Compendium of Dose Coefficients Based on ICRP Publication 60. *Ann. ICRP* **42**, 1–130 (2013).
5. Valentin, J. Guide for the Practical Application of the ICRP Human Respiratory Tract Model. *ICRP* **32**, (2002).
6. The Radiological Accident in Goiânia.
7. Naturally Occurring Radioactive Materials NORM - World Nuclear Association. https://world-nuclear.org/information-library/safety-and-security/radiation-and-health/naturally-occurring-radioactive-materials-norm.aspx.
8. Radiation Protection Services - Ionising Radiation and you. https://www.ukhsa-protectionservices.org.uk/radiationandyou/.
9. Ghiassi-nejad, M., Mortazavi, S. M. J., Cameron, J. R., Niroomand-rad, A. & Karam, P. A. Very high background radiation areas of Ramsar, Iran: preliminary biological studies. *Health Phys.* **82**, 87–93 (2002).
10. Nair, R. R. K. *et al.* Background radiation and cancer incidence in Kerala, India-Karanagappally cohort study. *Health Phys.* **96**, 55–66 (2009).
11. The Environmental Permitting (England and Wales) Regulations 2016. https://www.legislation.gov.uk/ukdsi/2016/9780111150184/schedule/23/data.htm?wrap=true.
12. Radioactive Wastes in the UK: A Summary of the 2010 Inventory.
13. Chernobyl: the true scale of the accident. https://www.who.int/news/item/05-09-2005-chernobyl-the-true-scale-of-the-accident.
14. The Roadway Safety Problem I US Department of Transportation. https://www.transportation.gov/NRSS/SafetyProblem.
15. Excess deaths associated with flu highest in 5 years. *GOV.UK* https://www.gov.uk/government/news/excess-deaths-associated-with-flu-highest-in-5-years (2023).
16. Provisional Mortality Statistics, Jan - Dec 2023 I Australian Bureau of Statistics. https://www.abs.gov.au/statistics/health/causes-death/provisional-mortality-statistics/latest-release (2024).

© The Editor(s) (if applicable) and The Author(s), under exclusive license to Springer Nature Switzerland AG 2025
J. Wain, *Ionising Radiation Protection*, Synthesis Lectures on Engineering, Science, and Technology, https://doi.org/10.1007/978-3-031-65525-8